The Palgrave Macmillan Animal Ethics Series

Series editors: **Andrew Linzey** and **Priscilla Cohn**

In recent years, there has been a growing interest in the ethics of our treatment of animals. Philosophers have led the way, and now a range of other scholars have followed, from historians to social scientists. From being a marginal issue, animals have become an emerging issue in ethics and in multidisciplinary inquiry. This series explores the challenges that Animal Ethics poses, both conceptually and practically, to traditional understandings of human–animal relations.

Specifically, the Series will:

- provide a range of key introductory and advanced texts that map out ethical positions on animals;
- publish pioneering work written by new, as well as accomplished, scholars; and
- produce texts from a variety of disciplines that are multidisciplinary in character or have multidisciplinary relevance.

Titles include:

Alasdair Cochrane
AN INTRODUCTION TO ANIMALS AND POLITICAL THEORY

Andrew Knight
THE COSTS AND BENEFITS OF ANIMAL EXPERIMENTS

Claire Molloy
POPULAR MEDIA AND ANIMAL ETHICS

Siobhan O'Sullivan
ANIMALS, EQUALITY AND DEMOCRACY

Thomas Ryan
SOCIAL WORK AND ANIMALS: A MORAL INTRODUCTION

Joan Schaffner
AN INTRODUCTION TO ANIMALS AND THE LAW

Forthcoming titles:

Aysha Akhtar
HUMANS AND ANIMALS: THE NEW PUBLIC HEALTH PARADIGM

Mark Bernstein
HUMAN–ANIMAL RELATIONS: THE OBLIGATION TO CARE

Eleonora Gullone
ANIMAL ABUSE AND HUMAN AGGRESSION

Alastair Harden
ANIMALS IN THE CLASSICAL WORLD: ETHICAL PERCEPTIONS

Lisa Johnson
POWER, KNOWLEDGE, ANIMALS

Kay Peggs
AN INTRODUCTION TO ANIMALS AND SOCIOLOGY

The Palgrave Macmillan Animal Ethics Series
Series Standing Order ISBN 978–0–230–57686–5 Hardback
978–0–230–57687–2 Paperback
(outside North America only)

You can receive future titles in this series as they are published by placing a standing order.
Please contact your bookseller or, in case of difficulty, write to us at the address below with
your name and address, the title of the series and one of the ISBNs quoted above.

Customer Services Department, Macmillan Distribution Ltd., Houndmills, Basingstoke,
Hampshire RG21 6XS, England

The Costs and Benefits of Animal Experiments

Andrew Knight
Oxford Centre for Animal Ethics, Oxford, England

First published 2011 by
PALGRAVE MACMILLAN

Palgrave Macmillan in the UK is an imprint of Macmillan Publishers Limited, registered in England, company number 785998, of Houndmills, Basingstoke, Hampshire RG21 6XS.

Palgrave Macmillan in the US is a division of St Martin's Press LLC, 175 Fifth Avenue, New York, NY 10010.

Palgrave Macmillan is the global academic imprint of the above companies and has companies and representatives throughout the world.

Palgrave® and Macmillan® are registered trademarks in the United States, the United Kingdom, Europe and other countries.

ISBN: 978–0–230–24392–7 hardback

This book is printed on paper suitable for recycling and made from fully managed and sustained forest sources. Logging, pulping and manufacturing processes are expected to conform to the environmental regulations of the country of origin.

A catalogue record for this book is available from the British Library.

Library of Congress Cataloging-in-Publication Data

Knight, Andrew, 1970–
    The costs and benefits of animal experiments / Andrew Knight.
        p. cm.
    Includes index.
        ISBN 978–0–230–24392–7 (hardback)
        1. Animal experimentation – Economic aspects. 2. Animal experimentation – Moral and ethical aspects. 3. Laboratory animals. I. Title.

HV4915.K64 2011
179′.4—dc22   2011007798

10 9 8 7 6 5 4 3 2 1
20 19 18 17 16 15 14 13 12 11

Printed and bound in Great Britain by
CPI Antony Rowe, Chippenham and Eastbourne

Contents

Tables

Figures

Abbreviations

ADME	absorption, distribution, metabolism, and excretion (see Glossary: *kinetics*)
AIDS	acquired immunodeficiency syndrome
AVA	Australian Veterinary Association
AVMA	American Veterinary Medical Association
AWA	US *Animal Welfare Act*
AWIC	Animal Welfare Information Center (US)
CA	chromosomal aberration
CAL	computer-aided learning
CCAC	Canadian Council on Animal Care
cDNA	complementary DNA (see Glossary)
χ^2	chi square (statistical) value (see Glossary: *chi square test*)
CI	confidence interval (see Glossary)
df	degrees of freedom (see Glossary)
EBM	evidence-based medicine (see Glossary)
EC	Commission of the European Communities
ECVAM	European Centre for the Validation of Alternative Methods
EPA	Environmental Protection Agency (US)
ESC	embryonic stem cell (see Glossary: *stem cell lines*)
EST	embryonic stem cell test
FDA	Food and Drug Administration (US)
FE	finite element analysis (an engineering technique)
fMRI	functional magnetic resonance imaging
GC	genotoxic carcinogen
GLP	good laboratory practice
GM	genetically modified (see Glossary)
GMP	good manufacturing practice
HCV	hepatitis C virus
hESC	human embryonic stem cell
HIV	human immunodeficiency virus
HPPC	Hans Popper Primate Center (Austria)
HPV	High Production Volume Challenge Program (US)
IACUC	institutional animal care and use committee
IARC	International Agency for Research on Cancer (World Health Organization)

IdMOC	Integrated Discrete Multiple Organ Co-culture system
ILSI	International Life Sciences Institute (US)
IRIS	Integrated Risk Information System (EPA)
LAL	limulus (horseshoe crab) amoebocyte lysate test
LGDW	laser-guided direct writing
LIF	leukaemia inhibitory factor
LLLT	low-level laser therapy
MEP	Member of the European Parliament
MHC	major histocompatibility complex
mRNA	messenger RNA
MTD	maximum tolerated dose (see Glossary)
NCI	National Cancer Institute (US)
NCRR	National Center for Research Resources (NIH, US)
NGC	non-genotoxic carcinogen
NHP	non-human primate
NIH	National Institutes of Health (US)
NOAEL	no observed adverse effect level (see Glossary)
NSAID	non-steroidal anti-inflammatory drug
NTP	National Toxicology Program (US)
NYBC	New York Blood Center
NZVA	New Zealand Veterinary Association
OECD	Organisation for Economic Co-operation and Development
OR	odds ratio (see Glossary)
PAB	polyclonal antibody
PBBK	physiology-based biokinetics (see Glossary: *-kinetics*)
PBPK	physiology-based pharmacokinetics (see Glossary: *-kinetics*)
PD	pharmacodynamics (see Glossary: *-dynamics*)
PET	positron emission tomography
PK	pharmacokinetics (see Glossary: *-kinetics*)
QA	quality assurance
QSAR	quantitative structure–activity relationship (see Glossary)
REACH	Registration, Evaluation and Authorisation of Chemicals testing programme (EC; see Glossary)
RP	relative predictivity
SAR	structure–activity relationship (see Glossary)
SC	stem cell (see Glossary: *stem cell lines*)
SHE	Syrian hamster embryo

SVM	School of Veterinary Medicine
TCR	T cell receptor
ToM	Theory of Mind (see Glossary)
USDA	United States Department of Agriculture
WHO	World Health Organization

Acknowledgements

The inspiration for this book was provided by Professor Andrew Linzey, Director of the Oxford Centre for Animal Ethics. I am very grateful for his encouragement, without which this book would not exist. I am also grateful for the contributions of the co-authors, editors, and numerous anonymous reviewers of my scientific publications which form the basis of much of this book. One of those co-authors was my partner, Jasmijn de Boo, who placed her research on hold to review this book. I am very grateful for her time and expertise, and for her tolerance of the temporary transformation of her partner into a single-minded wordsmith. Finally, I am grateful to Palgrave Macmillan's production team for their care and professionalism, without which this book would have remained a much rougher piece of work.

Excerpts and illustrations from the following publications appear in various chapters of this book, particularly where noted below, and are republished with the kind permission of the copyright holders.

De Boo J & Knight A (2008). Increasing the implementation of alternatives to laboratory animal use. *AATEX* 13(3): 109–17. Chapter 9.

Home Office (2010). *Statistics of Scientific Procedures on Living Animals: Great Britain 2009*. London: The Stationery Office. Figures 2.1–2.3.

Knight A (1999). Alternatives to harmful animal use in tertiary education. *Altern Lab Anim* 27(6): 967–74. Chapter 10.

Knight A (2007). Systematic reviews of animal experiments demonstrate poor human clinical and toxicological utility. *Altern Lab Anim* 35(6): 641–59. Chapters 2, 5–7, and 12.

Knight A (2007). The poor contribution of chimpanzee experiments to biomedical progress. *J Appl Anim Welf Sci* 10(4): 281–308. Chapter 5; Figures 5.1–5.4. Reprinted by permission of the publisher (Taylor & Francis Ltd, http://www.tandf.co.uk/journals).

Knight A (2007). The effectiveness of humane teaching methods in veterinary education. *ALTEX* 24(2): 91–109. Chapter 10; Tables 10.1–10.3.

Knight A (2008). 127 million non-human vertebrates used worldwide for scientific purposes in 2005. *Altern Lab Anim* 36(5): 494–6. Chapter 2.

Knight A (2008). Systematic reviews of animal experiments demonstrate poor contributions toward human healthcare. *Rev Recent Clin Trials* 3(2): 89–96. Chapters 5 and 7.

Knight A (2008). The beginning of the end for chimpanzee experiments? *Philos Ethics Humanit Med* 3: 16. Chapters 1, 4–5, and 12–13; Figures 5.1–5.4.

Knight A (2008). Non-animal methodologies within biomedical research and toxicity testing. *ALTEX* 25(3): 213–31. Chapter 8; Figure 8.1.

Knight A (2008). Advancing animal welfare standards within the veterinary profession. *REDVET: Revista electrónica de Veterinaria* 9(10B). Chapters 10–11.

Knight A, Bailey J & Balcombe J (2006). Animal carcinogenicity studies: 1. Poor human predictivity. *Altern Lab Anim* 34: 19–27. Chapter 6; Figure 6.1, Tables 6.2 and 6.4.

Knight A, Bailey J & Balcombe J (2006). Animal carcinogenicity studies: 2. Obstacles to extrapolation of data to humans. *Altern Lab Anim* 34: 29–38. Chapter 7; Figures 7.1–7.3 and Tables 7.1–7.2.

Knight A, Bailey J & Balcombe J (2006). Animal carcinogenicity studies: 3. Alternatives to the bioassay. *Altern Lab Anim* 34: 39–48. Chapter 8.

Knight A, Bailey J & Balcombe J (2006). Animal carcinogenicity studies: implications for the REACH system. *Altern Lab Anim* 34(Suppl. 1): 139–47. Chapter 6; Table 6.1.

Series Preface

This is a new book series for a new field of inquiry: Animal Ethics.

In recent years, there has been a growing interest in the ethics of our treatment of animals. Philosophers have led the way, and now a range of other scholars have followed, from historians to social scientists. From being a marginal issue, animals have become an emerging issue in ethics and in multidisciplinary inquiry.

In addition, a rethink of the status of animals has been fuelled by a range of scientific investigations which have revealed the complexity of animal sentiency, cognition, and awareness. The ethical implications of this new knowledge have yet to be properly evaluated, but it is becoming clear that the old view that animals are mere things, tools, machines, or commodities cannot be sustained ethically.

But it is not only philosophy and science that are putting animals on the agenda. Increasingly, in Europe and the United States, animals are becoming a political issue as political parties vie for the 'green' and 'animal' vote. In turn, political scientists are beginning to look again at the history of political thought in relation to animals, and historians are beginning to revisit the political history of animal protection.

As animals have grown as an issue of importance, so have there been more collaborative academic ventures leading to conference volumes, special journal issues, indeed new academic animal journals as well. Moreover, we have witnessed the growth of academic courses, as well as university posts, in Animal Ethics, Animal Welfare, Animal Rights, Animal Law, Animals and Philosophy, Human–Animal Studies, Critical Animal Studies, Animals and Society, Animals in Literature, Animals and Religion – tangible signs that a new academic discipline is emerging.

'Animal Ethics' is the new term for the academic exploration of the moral status of the non-human – an exploration that explicitly involves a focus on what we owe animals morally, and which also helps us to understand the influences – social, legal, cultural, religious, and political – that legitimate animal abuse. This series explores the challenges that Animal Ethics poses, both conceptually and practically, to traditional understandings of human–animal relations.

The series is needed for three reasons: (i) to provide the texts that will service the new university courses on animals; (ii) to support the increasing number of students studying and academics researching in animal-related fields; and (iii) because there is currently no book series that is a focus for multidisciplinary research in the field.

Specifically, the series will

- provide a range of key introductory and advanced texts that map out ethical positions on animals;
- publish pioneering work written by new, as well as accomplished, scholars; and
- produce texts from a variety of disciplines that are multidisciplinary in character or have multidisciplinary relevance.

The new Palgrave Macmillan Series on Animal Ethics is the result of a unique partnership between Palgrave Macmillan and the Ferrater Mora Oxford Centre for Animal Ethics. The series is an integral part of the mission of the Centre to put animals on the intellectual agenda by facilitating academic research and publication. The series is also a natural complement to one of the Centre's other major projects, the *Journal of Animal Ethics*. The Centre is an independent think-tank for the advancement of progressive thought about animals, and is the first centre of its kind in the world. It aims to demonstrate rigorous intellectual enquiry and the highest standards of scholarship. It strives to be a world-class centre of academic excellence in its field.

We invite academics to visit www.oxfordanimalethics.com, the Centre's website, and to contact us with new book proposals for the series.

ANDREW LINZEY AND PRISCILLA N. COHN
General Editors

1
Introduction

Legal implications of chimpanzee 'personhood'

On the basis of scientific argument that chimpanzees possess a *Theory of Mind** (ToM), and ought to be classified within the genus *Homo*, in 2007 advocates sought recognition under Austrian law of the *personhood* of a chimpanzee named Matthew Hiasl Pan (e.g. Balluch 2007). Success would mean that Matthew could no longer be considered property. He would become legally eligible for guardianship, on the basis that he was abducted as an infant, has been involuntarily confined in an alien environment for most of his lifetime, and has consequently been unable to fend for himself, or to safeguard his own interests. Through a guardian Matthew would be able to receive donations towards his living costs, and potentially even sue those responsible for his capture in West Africa in 1982 for acquired immunodeficiency syndrome (AIDS) and hepatitis research (Goodall 2006, Balluch 2007, Stafford 2007).

Matthew's case is highly controversial, and its legal and philosophical ramifications would be enormous, should it be upheld. At the time of writing, Matthew's advocates had not succeeded in the Austrian courts, and were appealing to the European Court of Human Rights. Whether or not they ultimately succeed, rapidly growing interest in such cases – including the publication of detailed legal foundations supporting the legislative personhood of chimpanzees (Wise 2000) – strongly suggests that Matthew's lawsuit will not be the last of its kind.

---

* Explanations of many technical terms are provided in the Glossary at the end of the book. These terms are *italicised* where they first occur, or where additional explanation is provided.

Moral implications of animal abilities

Matthew's case raises fascinating questions regarding the sensory, cognitive, and social characteristics necessary to confer a moral or legal right to protection from capture, social disruption, involuntary confinement, and *invasive* or life-threatening experimentation. It also raises questions about which species possess such characteristics, and to what degree.

Fundamental human rights, such as the rights to life, liberty, and freedom from torture, were first directly linked to the concept of *personhood* by the eighteenth-century German philosopher Immanuel Kant (Sturma 1999). Although such anthropocentrism has recently been challenged (e.g. White 2007, Benz-Schwarzburg & Knight 2011), classical concepts of personhood continue to rely on the possession of human-like psychological characteristics, such as consciousness and self-consciousness, the capacity to experience a wide range of emotional states, and the possession of key *cognitive* abilities, including those giving rise to culture, language, and *Theory of Mind*. The latter is a core consciousness-based capacity of human beings. Those possessing a ToM are considered able to ascribe mental or psychological states to themselves and others, including perceptual states such as seeing, as well as beliefs and desires (Premack & Woodruff 1978, Bischof-Köhler 2000).

In 2011 Benz-Schwarzburg and I reviewed in detail the scientific evidence for the existence of ToM in great apes, cetaceans, corvids, and other animal species (Benz-Schwarzburg & Knight 2011). In many respects chimpanzees such as Matthew possess the necessary mental characteristics; in other respects they might not. Combined evidence from multiple studies suggests they understand the intentions, goals, visual (and sometimes auditory) perception, and knowledge of other chimpanzees. The understanding that beliefs may be false (*false belief understanding*) demonstrates awareness of the difference between reality and mental representations, which is considered a core criterion for advanced ToMs (Onishi & Baillargeon 2005). Although evidence of such awareness in chimpanzees remains lacking to date, they seem able to understand others within a 'perception-goal psychology' (Call & Tomasello 2008, Kaminski *et al.* 2008, Krachun *et al.* 2009).

It is logically consistent to consider animals who possess such abilities as *non-human persons* who should be granted at least basic rights concordant with some of those granted to humans. One implication is that the moral boundary which ethical actors are obliged to respect is violated when such animals are used for a range of contemporary human purposes, including involuntary confinement and participation in biomedical research.

Despite morally important similarities, such species nevertheless differ from human persons (as they do from each other). However, it remains reasonable to conclude that they should be included in the community of moral consideration. After all, so-called *marginal human persons*, such as the very young, old, injured, or ill, who lack the full range of psychological and social characteristics and abilities exhibited by healthy human adults, are nevertheless valued as persons. They are valued as partially conscious, partially self-conscious, or partially autonomous beings, with unique personalities, and are accordingly granted human rights.

The Kantian foundation for human rights stems from the ideal of the rational person and the principle that every person is equally rational, self-conscious, and autonomous. However, the case for the equal application of human rights transcends this core idea. No matter how equal all humans actually *are*, all are *considered* equal in dignity and rights.

Utilitarian basis for animal experimentation

Current regulations governing animal experimentation fall far short of the moral consideration warranted by scientific advances in the understanding of key animal abilities and characteristics. Nevertheless, the interests of animals remain fundamentally important – at least in theory. The core principle underpinning animal experimentation regulation and policy is that the likely benefits of such research must outweigh its expected costs. Although considerable financial and human collateral costs do exist, the main costs are borne by the animals subjected to such research. And although such research may be directed at yielding benefits for animal species or the environment, the overwhelming majority is intended for human benefit, whether through the advancement of knowledge, through the development or toxicity testing of clinical interventions and consumer or industrial products, or through educational applications.

This *utilitarian* cost:benefit analysis underpins all fundamental regulation governing animal experimentation. *Directive 2010/63/EU on the protection of animals used for scientific purposes*, which directs such animal use in all EU member states, asserts that it is 'essential, both on moral and scientific grounds, to ensure that each use of an animal is carefully evaluated as to the scientific or educational validity, usefulness and relevance of the expected result of that use. The likely harm to the animal should be balanced against the expected benefits of the project' (EU 2010).

However, the contemporary widespread reliance on animal models in biomedical research and toxicity testing is heavily dependent on assumptions of human utility – and, in particular, of reasonable *predictivity* for human outcomes. Surprisingly, these assumptions have rarely been verified or, indeed, subjected to rigorous scrutiny of any kind.

Such untested or unfounded assumptions about the human utility of animal models have rendered utilitarian arguments for and against their use largely speculative to date.

Purpose of this book

To judge the merits of animal experimentation overall, it is essential to have reliable information about the magnitude of laboratory animal use, the probable harms inflicted on animals, the human benefits realised, and the potential offered by alternative research, testing, and educational methodologies. Accordingly, these key topics are investigated in the corresponding parts of this book:

Part I (Chapters 2–4)	Animal Costs
Part II (Chapters 5–7)	Human Benefits
Part III (Chapters 8–9)	Alternative Strategies
Part IV (Chapters 10–11)	Educational Animal Use and Student Impacts

In addition, in each of the three fields of fundamental and clinically applied biomedical research, toxicity testing, and biomedical education, a key example is examined in greater detail.

Invasive chimpanzee research

Chimpanzees are the species most closely related to humans, and consequently most likely to be generally predictive of human outcomes when used in research aimed at the development of human clinical interventions. However, their advanced cognitive, psychological, and social characteristics also raise exceptional animal welfare and ethical concerns when they are confined in research laboratories and subjected to invasive procedures. Both the likely benefits and the ethical costs are potentially maximised by such research. These costs and benefits are examined in detail in Chapters 4, 5, and 12.

Animal carcinogenicity testing

In the field of toxicity testing, the accurate identification of previously unknown human *carcinogens* to which workers, consumers, or patients may be exposed offers greater potential public health benefits than the identification of all other toxins combined – by a considerable margin. Accordingly, the accurate identification of such carcinogens is crucial. However, due to a lack of human exposure data, carcinogen testing has traditionally relied heavily on animal studies. Chapter 6 examines the utility of such studies in accurately predicting human carcinogenicity, and in deriving hazard classifications for the regulation of human exposure.

Veterinary education

Veterinarians must be able to perform a variety of clinical and surgical procedures on animal patients, including euthanasia, and must be familiar with the clinical signs of animal diseases. Accordingly, the justifications for invasive animal use are stronger in veterinary education than in virtually any other educational discipline. To critically assess the necessity of invasive animal use in education, it is therefore instructive to examine the case of veterinary education closely.

Participation in such animal use may also profoundly affect the development of attitudes towards animal welfare – which are fundamentally important in the case of veterinarians. However, such attitudinal impacts have been relatively under-studied to date. These topics are examined in detail in Chapters 10 and 11.

Conclusions and policy recommendations

Animal experimentation is arguably the single most contentious issue in the wider debate concerning the rights of humans to use animals. However, detailed critical review indicates that, although uncertainties remain, sufficient evidence now exists to draw some key conclusions about the overall costs to animals, and benefits to humans, of invasive animal experimentation.

Examining and weighing these costs and benefits leads to important recommendations for the ethical oversight of scientific animal use, for the scientific *validation* and acceptance of both animal and non-animal experimental models, for the implementation of alternative research, testing, and educational strategies, and for the development of associated policy and regulation.

Appropriate recommendations are summarised, alongside an overview of existing regulation governing laboratory animal use in Europe and the US, in:

Part V (Chapters 12–13)   Conclusions and Policy Recommendations

Use of this book

This book uses some technical language. To assist readers unfamiliar with them, many technical terms are explained in the Glossary at the end of the book. These terms are *italicised* where they first occur, or where further explanations are provided in the text.

Readers should be aware that approximations of summated totals and their numerical components in figures, tables, and the text are all correctly rounded from original figures, as are corresponding percentages. However, such rounded approximations do not always summate perfectly.

Concluding summaries are provided in all of the following chapters with the exception of those in Part V, which summarise the remainder of this book. Full colour versions of the figures may be viewed at www. palgrave.com/animalexperiments.

Intended readership

This book is intended to serve all who are interested in the scientific and educational utility of laboratory animal use, and in alternative research, testing, and educational strategies. It may be of use to scientists and educators working with animals, or developing alternatives; to policy-makers, including regulatory agencies and legislators; to chemical and pharmaceutical companies and consumer product manufacturers, who are increasingly required to provide toxicity data on their products; to both undergraduates and postgraduates studying the ethical issues surrounding animal experimentation; to bioethicists and philosophers concerned with animal issues; and to members and supporters of organisations promoting scientific animal use, the protection of animals, and patient and consumer safety.

Part I
Animal Costs

2
Global Laboratory Animal Use

Most animal experimentation is intended to benefit human beings. To assess the merits of such research, one must weigh the likely human benefits against the probable costs to the animals involved. The potential offered by alternative methodologies should also be considered. In any determination of the overall impacts experienced by research animals, and the likely future impacts, reasonably accurate assessments of the number of animals used, both globally and in specific regions and countries, are important, as are trends in animal use. Unfortunately, the accuracy of animal use estimates has been significantly impeded by wide international variation in reporting standards. The US, for example, excludes mice, rats, birds, fish, reptiles, and amphibians from official statistics (USDA 2005). However, independent calculations reveal that these species jointly make up well over 90 per cent of the US total (Taylor *et al.* 2008). Many countries fail to record or publish animal use statistics at all, and of those that do, most record only live animal use, excluding the substantial number of animals killed prior to procedures such as the harvesting of organs or other tissues for scientific purposes. Despite such limitations, it remains clear that many millions of animals are used worldwide each year, and that certain trends are substantially increasing laboratory animal use.

Global animal use

Several estimates of laboratory animal use globally, or in major world regions, have been published to date. Some were recently summarised by Taylor and colleagues (2008). These included 100–200 million used globally in 1970 (Van Zutphen 2001), 60–85 million used globally in 1993 (Rowan 1993), 42 million used in 17 countries in 1998

(Orlans 1998), 28 million used in 21 mainly European countries in 2000 (Orlans 2000), over 29 million used in European countries and North America in 2004 (Reinhardt & Reinhardt 2006), and 50–100 million used globally in 2005 (Nuffield Council on Bioethics 2005). However, most of these estimates were based on extrapolation significantly beyond a smaller pool of known national figures or relied on expert opinions, and the widely varying results aptly demonstrate the resultant uncertainties.

Accordingly, Taylor and colleagues sought to provide a large-scale evidence-based estimate of global laboratory animal use. After adjusting official data for 37 countries to match EU definitions of animals, experimental procedures, and other relevant criteria, they estimated that a total of 50.4 million animals were used by these countries in 2005.

In addition, by demonstrating a highly significant positive linear correlation between animal use in these countries and animal study publication rates the following year, they were able to estimate that 7.9 million additional animals were used in 142 remaining countries for which only publication rates were available. Publication rates of animal studies were obtained by surveying the published scientific literature. This step allowed the inclusion of all nations with a population greater than 200,000.

Hence, Taylor and colleagues estimated that a total of 58.3 million living non-human vertebrates were subjected to fundamental or medically applied biomedical research, toxicity testing, or educational use in these 179 countries in 2005.

However, several important categories of animal use were initially excluded, including animals killed for the provision of experimental tissues, used to maintain established *genetically modified* (GM) strains, or bred for laboratory use but killed as surplus to requirements. Such categories also raise significant bioethical concerns, and are important in considering the merits of laboratory animal use. When these additional categories were included, the estimate increased to a total of 115.3 million non-human vertebrates used worldwide.

In calculating these additional totals, however, Taylor and colleagues relied on unweighted, or *arithmetic means*. When *weighted means* were used, which more accurately reflected the proportional contribution of each country to the overall totals, some 68.6 million animals were calculated to fall into these additional categories. This resulted in a grand total of 126.9 million non-human vertebrates used worldwide in 2005 (Knight 2008a). At the time of writing this remained the most accurate recent, evidence-based estimate of global laboratory animal numbers.

Due to several factors, however, these estimates were considered to be highly conservative. As identified by Taylor and colleagues, for example, their estimate of 17.3 million living vertebrates used in the US is very significantly lower than a 2000 governmental estimate of 31–156 million, which was based on extrapolation from the results of a survey of only 50 of 2,000 research institutions (USDA 2000). Furthermore, Taylor and colleagues' estimates excluded several other categories of ethical concern, such as the use of invertebrate species now understood to have advanced capacity for suffering, including certain *cephalopod molluscs* and *decapod crustaceans*, and the use of advanced foetal developmental stages, which affects both the foetuses and the *dams* from which they are sourced.

Greatest animal users

As estimated by Taylor and colleagues, the national regions making greatest use of laboratory animals in 2005 are listed in Table 2.1. In some cases very marked disparities between estimated totals and those reported by government agencies demonstrate the lack of international consistency of the latter. The US, for example, was the greatest user of laboratory animals, using an estimated 17.3 million. However, the exclusion of mice, rats, birds, fish, reptiles, and amphibians resulted in an official US total of only 1.2 million. Great Britain, on the other hand, included foetal forms, the maintenance of GM breeding colonies, and some tissue harvesting in its official total of 2.8 million – none of which was included in the standardised EU definition of animal use applied by Taylor and colleagues. Its total was therefore adjusted downwards to 1.9 million.

The EU was the only region publishing harmonised statistics on laboratory animal use for its member states. Just over 12.0 million animals were used in the 27 EU member states in 2008 – the latest reporting period at the time of writing (other than France, which provided figures for 2007) (EC 2010a).

Table 2.1   The 12 national regions using more than 1 million laboratory animals in 2005

	Region	Animals used in 2005
1	US	17,317,147
2	Japan	11,154,961
3	China	2,975,122
4	Australia	2,389,813
5	France[a]	2,325,398
6	Canada	2,316,281
7	Great Britain	1,874,207
8	Germany	1,822,424
9	Taiwan	1,237,337
10	Brazil	1,169,517
11	Thailand	1,059,355
12	Norway	1,000,426

[a] France supplied figures for 2004.
Note: EU definitions of 'animal' were applied.

Data source: Taylor *et al.* (2008).

Increases in animal use

In recent years laboratory animal use has increased in many countries. Two major factors appear to be responsible for this trend, at least in Europe and the US. These are the increased production and use of GM animals, and the initiation of large-scale chemical testing programmes, which are expected to require unprecedented numbers of animals.

Use of genetically modified animals

Previous steady declines in laboratory animal use have reversed in some countries in recent years, mostly as a result of dramatic increases in the use of GM animals. The production of GM animals also necessitates substantial breeding, further increasing animal numbers. In Great Britain, for example, after a steady and significant reduction from 1976 onwards, the figures stabilised during the late 1990s, and then the trend reversed. Just over 3.6 million regulated scientific procedures (i.e. those likely to cause pain, suffering, distress, or lasting harm to living vertebrates and one species of octopus) were initiated in 2009 – the second highest total since the current method of recording was introduced in 1988, more than two decades ago. A 1.0 per cent reduction from the 2008 maximum was preceded by seven successive annual increases, and

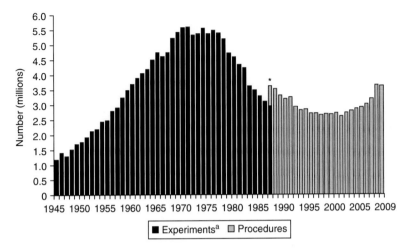

Figure 2.1 Experiments or procedures using animals begun annually in Great Britain, 1945–2009

ᵃDefined as experiments under the *Cruelty to Animals Act* (1876) or as scientific procedures under the *Animals (Scientific Procedures) Act* (1986).

*The 1987 total includes experiments under the 1876 Act as well as procedures under the 1986 Act.

Source: Home Office (2010). Reproduced under the terms of the Click-Use Licence.

the total number of procedures remained a third higher (approximately 904,800, or 33.3 per cent) than in 2000 (Figure 2.1) (Home Office 2010, Knight 2010).

The proportion of *genetically altered* animals used has been steadily rising since at least 1995, and in 2009 exceeded the number of normal animals used for the first time: in 2009, 52.4 per cent of procedures involved animals that were genetically altered. These comprised 382,900 procedures (10.6 per cent) using animals with harmful genetic mutations, and 1.5 million procedures (41.8 per cent) that used GM animals. The number of procedures using genetically normal animals was 1.7 million (47.6 per cent; Figure 2.2) (Home Office 2010).

Breeding for the production of harmful mutant and GM animals accounts for a sizeable proportion of total animal use, which has been increasing over time (Figure 2.3). In 2009 such breeding procedures accounted for 42.4 per cent (1.5 million) of the total of just over 3.6 million procedures initiated in Great Britain – an increase of 143,000 (10.3 per cent) over the preceding year (Home Office 2010). Indeed, if breeding procedures were excluded, then the total number

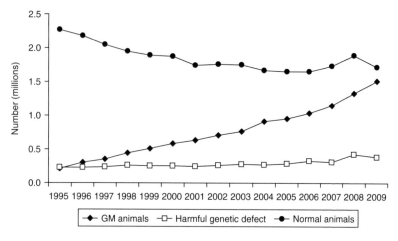

Figure 2.2   Procedures by genetic status of animal in Great Britain, 1995–2009
Note: GM, genetically modified.
Source: Home Office (2010). Reproduced under the terms of the Click-Use Licence.

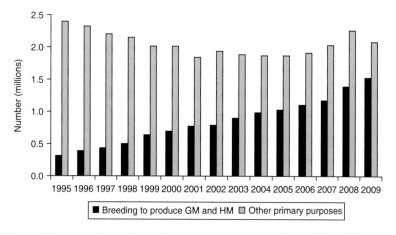

Figure 2.3   Comparison of breeding to produce genetically modified (GM) animals and animals with harmful mutations (HM) with other primary purposes in Great Britain, 1995–2009
Source: Home Office (2010). Reproduced under the terms of the Click-Use Licence.

of procedures would actually have decreased by 179,600 (7.9 per cent) when compared with 2008, to 2.1 million. Of these remaining 2.1 million procedures, 622,800 (29.9 per cent) used genetically altered animals.

Substantial increases in GM animal use have similarly been recorded in Germany (Sauer *et al.* 2005) and Switzerland (HSUS 2007), where total animal use is also increasing (Rusche 2003, HSUS 2007).

Chemical testing programmes

Large-scale chemical testing programmes are also important drivers of recent and probable substantial future increases in laboratory animal use (Combes *et al.* 2004, Green & Goldberg 2004). These programmes are intended to bridge knowledge gaps regarding the toxicity of chemicals that are produced in or imported into Europe or the US in particularly high quantities, or that otherwise raise special concerns (Combes *et al.* 2004, Green & Goldberg 2004). Such programmes follow increasing public, political, and regulatory concern about the potential toxicity of a wide range of environmental, occupational, and consumer chemicals, and are likely to result in the use of unprecedented numbers of animals.

Prominent among these programmes is the 2003 Commission of the European Communities (EC) proposal for the *Registration, Evaluation and Authorisation of Chemicals* (REACH), which entered into force on 1 June 2007. REACH aims to assess the toxicity of some 30,000 chemicals produced in or imported into the EU in quantities in excess of 1 tonne annually.

The testing requirements of REACH are unprecedented. Reproductive and developmental toxicity data, for example, are required for all substances manufactured or imported in quantities exceeding 10 tonnes annually. It has been estimated that reliance on traditional whole animal testing could require the use of almost 22 million vertebrates, at a cost of up to several hundred thousand dollars per registered substance – although total animal numbers may reportedly be reduced to 10 million or fewer if proprietary in-house toxicity data and appropriate non-animal testing and evaluation methodologies are utilised (Scialli 2008).

In the US, two noteworthy programmes have been managed by the Environmental Protection Agency (EPA) since 1998. The *High Production Volume* (HPV) *Challenge Program* targets chemicals produced or imported in quantities in excess of 1 million pounds annually. As of 2007, data on more than 2,200 chemicals had been supplied or were being gathered (EPA 2010a). In 2006, Sandusky and colleagues reported that the HPV programme had already subjected over 150,000 animals to chemical tests (Sandusky *et al.* 2006).

The Endocrine Disrupting Chemicals Research Program aims to assess dietary and environmental chemicals with the potential to disrupt normal hormonal regulation in humans and wildlife. In 1998, the EPA's Endocrine Disruptor Screening and Testing Advisory Committee recommended a tiered testing approach for the evaluation of oestrogen-, androgen-, and thyroid hormone-related effects of some 87,000 commercial chemicals and environmental contaminants (Charles 2004).

The logistical challenges incurred through reliance on traditional *bioassays* (animal tests) to meet such unprecedented testing demands are aptly demonstrated by the rodent carcinogenicity bioassay. This assay takes upwards of two years to produce results of demonstrably poor human specificity (Knight *et al.* 2006a), at an average cost of €780,000 (Fleischer 2007, see also Combes *et al.* 2007). Unsurprisingly, by 1998, only some 2,000 (2.7 per cent) of the 75,000 industrial chemicals then in use and listed in the EPA Toxic Substances Control Act Inventory had been tested for carcinogenicity (Epstein 1998). The resources consumed by testing just these 2.7 per cent of industrial chemicals included hundreds of millions of US dollars (Conolly *et al.* 1988), millions of skilled personnel hours (Gold *et al.* 1999), and millions of animal lives (Monro and MacDonald 1998, Gold *et al.* 1999).

As noted by Bremer and colleagues (2007), requiring *in vivo* testing for every adverse effect in these high-throughput chemical testing programmes would exceed the capacity of available scientific facilities and expertise, would result in unacceptably high *false positive rates*, and would, in all likelihood, endanger the success of these programmes.

Such considerations have generated a strong interest in alternative testing strategies. This interest was increased when challenging timelines for phasing out animal testing of cosmetics after certain chronological deadlines, as well as marketing bans on cosmetics tested *in vivo*, were imposed in the EU under the seventh *Amending Directive 2003/15/EC to Cosmetics Directive 76/768/EEC* (Combes *et al.* 2007, Lilienblum *et al.* 2008).

Summary

The most accurate evidence-based estimates of global laboratory animal numbers currently available describe animal use in 2005. Around 58.3 million living non-human vertebrates were subjected to fundamental or medically applied biomedical research, toxicity testing, or educational use, and 68.6 million more were estimated to have been killed for the provision of experimental tissues or as surplus to requirements, or used to maintain established GM strains, none of which were included within EU standardised figures. This resulted in a total of approximately 126.9 million non-human vertebrates used worldwide in 2005 (Knight

2008a). The accuracy of these figures, and particularly the 68.6 million component, is markedly limited by lack of published animal use statistics. Nevertheless, for several reasons these estimates are considered to be highly conservative.

After applying EU definitions of animal use, the top three users of laboratory animals in 2005 were the US (approximately 17.3 million), Japan (approximately 11.2 million), and China (approximately 3.0 million). Australia, Canada, and several European countries also ranked high (Taylor *et al.* 2008).

Both globally and in most countries that publish statistics, laboratory animal use is increasing due to increased use and production of GM animals and the implementation – particularly in Europe and the US – of large-scale chemical testing programmes.

Finally, it must be remembered that these estimates exclude several other categories that raise ethical concerns, including the use of invertebrate species such as certain cephalopod molluscs and decapod crustaceans known to possess advanced capacities for suffering, and the use of foetuses in advanced stages of development.

3

Types of Laboratory Animal Use

The nature and magnitude of impacts experienced by laboratory animals depend on a wide range of factors. These typically include level of invasiveness of procedures; neurological and cognitive capacity, and developmental stage, of the animals used; extent of *analgesic* (pain-killer) and anaesthetic use; degree of domestication of the species involved; stresses involved in the capture, transportation, and relocation of wild-sourced animals (such as some primates); level of environmental enrichment provided in laboratory housing; and amount of social contact provided for social species.

The extent to which such factors are described in annual animal experimentation statistics (in countries where these statistics exist at all) is variable and incomplete. However, through examination of certain world regions many of the more important factors start to become clear.

Species used

The main species used, their origins, and types of use are indicated by examination of the 27 member states of the EU – which constitute the only major world region publishing harmonised statistics. EU statistics are currently published every three years, and at the time of writing the most recent statistics described laboratory animal use in 2008 – with the exception of France, which provided figures for 2007 (EC 2010a & 2010b).

Rodents and rabbits jointly made up 82.2 per cent of the total number of animals used. In this group, mice and rats accounted for most of the animals, comprising 59.4 per cent and 17.7 per cent respectively of the overall total. The second-largest group – at 9.6 per cent of the total – was made up of *poikilotherms* (cold-blooded organisms – which in this case and laboratory animal statistics generally refers to fish, amphibians, and reptiles). The third-most common animals used were birds, which represented 6.4 per cent of the total (EC 2010b) (Figure 3.1).

The group consisting of horses, donkeys, and crossbreeds (*Perissodactyla*) and pigs, goats, sheep, and cattle (*Artiodactyla*) represented only 1.4 per cent of the total. Carnivores made up only 0.3 per cent of the total, and non-human primates (NHPs) only 0.1 per cent. No great apes were used in 2008, or in 2005 and 2002 – the preceding reported years (EC 2010b).

Examination of the reports for 1996, 1999, 2002, 2005, and 2008 indicated reasonably close consistency over time in the main classes of animals used, even as the EU expanded from 15 to 27 member states reporting by 2008 (Table 3.1).

The percentages of rodents and rabbits combined fluctuated around 80 per cent. For poikilotherms, the proportions ranged from 6.6 per cent in 1999, to around 10–11 per cent in 1996 and 2008, and increased to just over 15 per cent in 2002 and 2005. The proportion of birds used rose steadily from 3.9 per cent in 1996, to 6.4 per cent in 2008, and the proportion of Artiodactyla and Perissodactyla similarly rose from 1.0 to 1.4 per cent during the same period.

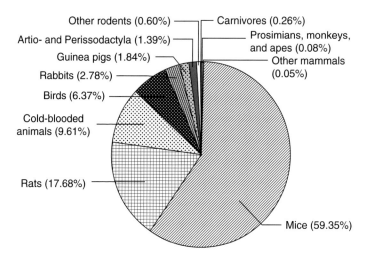

Figure 3.1 Proportions of laboratory animals used in the 27 EU member states, 2008[a]

[a] France supplied data for 2007.

Data source: EC (2010b).

Table 3.1   Proportions of main classes of animals used in EU member states, 1996–2008 (%)

Class of species	1996[a]	1999[b]	2002[c]	2005[d]	2008[e]
Rodents and rabbits	83.8	87.0	78.1	77.8	82.2
Cold-blooded animals	10.9	6.6	15.4	15.1	9.6
Birds	3.9	4.7	5.0	5.4	6.4
Artiodactyla and Perissodactyla	1.0	1.3	1.2	1.2	1.4

[a] Fourteen member states reporting for 1996, plus France reporting for 1997.
[b] Fifteen member states reporting for 1999, including France.
[c] Fourteen member states reporting for 2002, plus France reporting for 2001.
[d] Twenty-four member states reporting for 2005, plus France reporting for 2004.
[e] Twenty-six member states reporting for 2008, plus France reporting for 2007.

Data sources: EC (1999, 2003a, 2005, 2007 & 2010b).

Animal sourcing

Of the 9.9 million instances of animal use in EU countries in 2008 for which animal origins were reported, 8.3 million animals (84.4 per cent) originated from registered breeding or supplying establishments in the reporting country. However, 20–30 per cent of cats, dogs, and ferrets used, and around 45 per cent of *Old World monkeys*, were of non-European origin. As in previous years, no apes of any kind were used in 2008 (EC 2010b).

The origins of most species were similar to those observed in previous reported years. However, the *prosimians* used were all of European origin for the first time in 2005, and this remained the case in 2008. A similar trend was observed for *New World monkeys*, almost all of which originated from Europe. Finally, although the proportion of Old World monkeys from non-European origins remained very large, it decreased from around 60 per cent in 2005 to 45 per cent in 2008. The proportion of cats of European origin increased, whereas the proportions of dogs and ferrets of non-European origin remained unchanged from 2005 (EC 2010b).

Categories of use

The largest single category of use was fundamental biological studies (as distinct from those of an applied nature), which accounted for 38.1 per cent of all procedures. However, when research and development for clinical interventions (22.8 per cent) is added to the production and quality control of such interventions intended for human use (10.9 per cent),

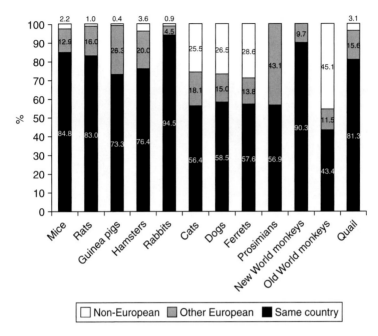

Figure 3.2 Countries of origin for laboratory animals used in the 27 EU member states, 2008[a]

[a] France supplied data for 2007.
Note: Only species whose origins had to be reported are included. 'Other European' countries were EC member states, or members of the Council of Europe, which, at the beginning of the reporting period, were parties to the *European Convention for the Protection of Vertebrate Animals Used for Experimental and Other Scientific Purposes (ETS 123)*. No apes were used in 2008. Some animals counted separately were, in fact, the same animals used repeatedly. However, other than for cats, dogs, and New World and Old World monkeys, the proportion of such animals was very small, and for all species it remained a minority.
Data source: EC (2010b).

and to toxicological and other safety evaluation (8.7 per cent), it can be seen that experiments focused on the development, production, or safety evaluation of clinical interventions and other products made up 42.4 per cent – the largest category overall. With the exception of a small proportion being developed and tested for veterinary use, almost all of these were intended for human applications (Figure 3.3; EC 2010b).

The most significant changes in comparison with 2005 (the previous reported year) included a sharp decline from 3.7 million (30.9 per cent) to 2.7 million (22.8 per cent) in the number of animals used for research and development for human medicine, dentistry, and veterinary medicine.

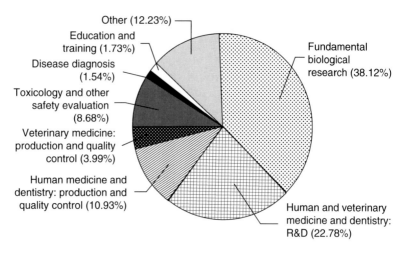

Figure 3.3   Categories of scientific animal use in the 27 EU member states, 2008[a]

[a] France supplied data for 2007.

Note: 'Other' purposes included virology, the production of monoclonal and polyclonal antibodies, physiology of foetal–maternal interaction in mouse gene transgenesis, onco-logical treatment, pharmaceutical R&D, combined drug testing, and genome studies.

Data source: EC (2010b).

This included a reduction of more than 800,000 poikilotherms. On the other hand, the number of animals used for fundamental biological research increased from 4.0 million (33.3 per cent) to 4.6 million (38.1 per cent), and the proportion of animals used for 'other' purposes also increased, from 8.1 per cent to 12.2 per cent (EC 2007 & 2010b).

Substantial increases occurred in the use of mice and rabbits for the production and quality control of human clinical interventions, and in the use of mice, pigs, and birds for fundamental biological research and 'other' experiments. Several European member states confirmed that the increase in the use of mice had resulted from the new research pos-sibilities offered by *transgenic* species (EC 2010a).

Procedural invasiveness

At the time of writing, procedural invasiveness was not directly indi-cated in EU reports. However, some figures were available for Australia and Canada, which were the fourth- and sixth-largest national users of laboratory animals respectively in 2005 (Table 2.1). Figures were also available in reports from the US (the leading user of laboratory animals); however, the exclusion of mice, rats, and birds from US reports effectively excludes the overwhelming majority of animal subjects, severely limiting the utility of the resultant figures.

Australia

In Australia, figures indicating procedural invasiveness were available for 2005–8 at the time of writing (Figure 3.4; HRA 2010). Unfortunately, data for just under half the Australian states and territories were available; however, these included New South Wales and Victoria, which can be assumed to be the greatest users of laboratory animals on the basis of state population sizes and earlier years for which usage figures were available.

To harmonise with EU definitions of scientific animal use, a large number of observational studies involving only minor interference have been excluded from consideration. Where states did not specify categories of animal use at all, their contributions were entirely excluded to ensure that no observational studies were inadvertently included. Hence, the resultant totals of definitely non-observational animal use (Figure 3.4) were very significantly lower than the total recorded use for each of these years.

In considering these results, it should be noted that *non-recovery procedures* result in the death of the subject and that the production of GM strains involves minor and major physiological challenges and surgical procedures. Their production is also an inherently inefficient process, frequently resulting in a high proportion of discarded animals, with the welfare of the survivors more likely to be adversely affected than for non-GM strains (Thon *et al.* 2002).

Procedures that can be considered *markedly invasive* include those resulting in death (whether or not the animals were conscious), surgical procedures (excluding minor operative procedures), major physiological challenges, and the production of GM animals. Jointly, such procedures were recorded as fluctuating between 23.3 and 39.6 per cent of all definitely non-observational animal use recorded during this four-year period. However, lack of reporting detail in states and territories means these results can be considered only an approximation of Australian animal use.

Canada

In Canada, animal use statistics have been provided to the Canadian Council on Animal Care (CCAC) by participating institutions since

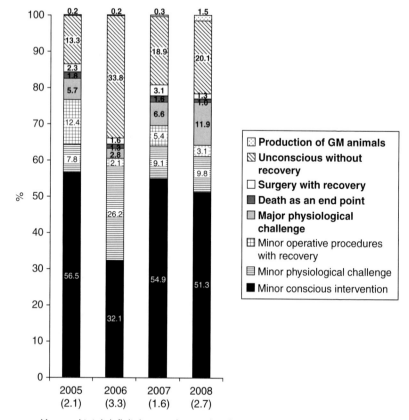

Figure 3.4 Invasiveness of definitely non-observational Australian laboratory animal use, 2005–8

Note: Large numbers of definitely or possibly observational studies have been excluded. Markedly invasive procedures are in **boldface**.

Data source: HRA (2010), who sourced their data from state government reports.

1975 (CCAC 2009). Since 1996, these have been provided in a consistent format. Invasiveness of use is indicated by Figure 3.5 and Table 3.2.

Markedly invasive procedures – those in category of invasiveness D or E – increased from a minimum of 28.8 per cent in 1996 to a maximum of 43.9 per cent in 2007, and then decreased slightly to 39.6 per cent in 2008. This range is slightly higher than the 23–40 per cent approximations recorded for Australia, but is probably significantly

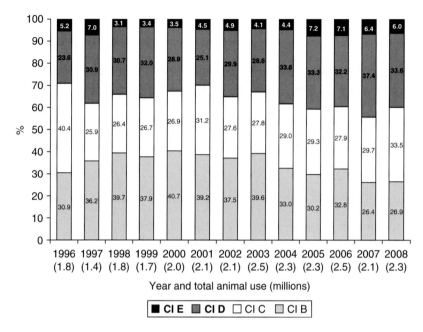

Figure 3.5   Invasiveness of Canadian laboratory animal use, 1996–2008

Note: Categories of invasiveness (CIs) are defined in Table 3.2. Markedly invasive procedures are in **boldface**.

Data source: CCAC (2009).

Table 3.2   Category of invasiveness (CI) of animal procedures in Canada

CI	Experiments cause
E	Severe pain near, at, or above the pain tolerance threshold of unanaesthetised conscious animals
D	Moderate to severe distress or discomfort
C	Minor stress or pain of short duration
B	Little or no discomfort or stress

Source: CCAC (2009).

more accurate due to the lack of unspecified procedures in the Canadian figures.

Examination of Canadian animal use also reveals a trend towards increased invasiveness of procedures over time. To some extent this may have resulted from increased understanding among researchers about animal capacities for suffering, more accurate monitoring of variables indicative of pain and stress, and more realistic classification of experiments. It may, however, also indicate a true increase in procedural invasiveness over time.

Anaesthetic use

At the time of writing, anaesthetic use was not directly indicated in EU reports, or those of Australia and Canada. It was, however, indicated in the reports of Great Britain – the seventh-largest animal user in 2005 (Table 2.1) – which has provided harmonised figures since 1988 (Figure 3.6) (Home Office 2010).

In 2009, 2.4 million British procedures (66.7 per cent of the overall total) did not use any form of anaesthesia – approximately 36,000 (1.5 per cent) more than in 2008. From 1988 to 2009 the proportion of such procedures fluctuated between approximately 59 per cent and 69 per cent.

General anaesthesia was provided throughout or at the end of *terminal procedures* in approximately 342,300 cases (9.5 per cent of the overall total) in 2009. Such anaesthetic use in association with terminal procedures has decreased from around 19 per cent in 1988.

Other forms of anaesthesia (local, regional, or general anaesthesia with recovery) were provided in approximately 863,400 procedures in 2009 (23.9 per cent of the overall total). The proportion of such procedures increased from around 17 per cent in 1988 to around 29 per cent in 2007, although it decreased thereafter. Around 243,300 procedures (6.7 per cent of the overall total) involved local anaesthesia in 2009, and approximately 620,000 (17.1 per cent of the overall total) involved general anaesthesia with recovery.

The use of *neuromuscular blocking agents* was also recorded during 3,600 procedures in 2009, all of which also used general anaesthesia.

Summary

Several conclusions can be drawn from these reports. On the basis of the latest EU reports (EC 2010a & 2010b), it is clear that the largest groups of animals used are mice (around 59 per cent) and rats (around 18 per cent). Next are poikilotherms, which constitute around 10 per cent of laboratory animals, and birds, which constitute around 6 per cent.

At least in the EU, most laboratory animals originate from breeding or supplying establishments in the same country, with the main

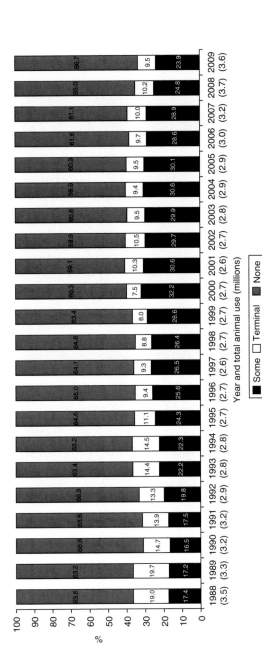

Figure 3.6 Proportions of animals undergoing scientific procedures in Great Britain provided with no, some, or terminal anaesthesia, 1988–2009

Note: Anaesthesia provided was local, regional, or general. Terminal anaesthesia was provided either at the end of the procedure or for the whole of the procedure concerned.

Data source: Home Office (2010).

exceptions being 20–30 per cent of cats, dogs, and ferrets and around 45 per cent of Old World monkeys, which originated from non-European countries.

Around 42 per cent of all EU experiments are focused on the development, production, or safety evaluation of clinical interventions and other products, almost all of which are intended for human use. The next-largest category is biological studies of a fundamental, rather than applied, nature, which make up around 38 per cent of all animal procedures.

On the basis of Canadian figures, the proportion of markedly invasive procedures has ranged from around 29 per cent to 44 per cent over the past decade, with just under 40 per cent recorded in 2008 – the most recent year for which data were available at the time of writing. Such procedures may be becoming more prevalent over time (CCAC 2009).

And on the basis of the British figures over the past 20 years, procedures not using any kind of anaesthetic fluctuated between approximately 59 and 69 per cent of recorded totals, with just under 67 per cent recorded in 2009 – the most recent year for which data are available. The animals in this category increased by 1.5 per cent compared with the previous year. General anaesthesia was provided throughout or at the end of terminal procedures in around 9 per cent of procedures overall, decreasing from 19 per cent in 1988. The proportion of procedures involving other forms of anaesthesia (local, regional, or general anaesthesia with recovery) increased from around 17 per cent in 1988 to 29 per cent in 2007, before decreasing to 24 per cent in 2009. Of these, just under 72 per cent involved general anaesthesia with recovery (Home Office 2010).

To further assess animal impacts it would be helpful to know the levels of analgesic use, of correlation between markedly invasive procedures and anaesthetic or analgesic use, and of environmental enrichment and socialisation opportunities. Unfortunately, such information remains largely unreported.

To assess the cumulative effect of adverse impacts during an animal's lifetime, which would provide a truer indication of animal welfare status, such assessment and recording would need to be ongoing and linked to each animal. For the most accurate assessment of welfare, other relevant factors such as husbandry standards and health and nutritional status would also need to be considered. However, such lifetime recording and assessment is extremely rare.

4
Impacts on Laboratory Animals

Stress is 'the effect produced by external (i.e., physical or environmental) events or internal (i.e., physiologic or psychologic) factors ... which induce an alteration in an animal's biologic equilibrium' (i.e. state of *homeostasis*). *Stressors* are adverse stimuli resulting in stress, and *distress* is 'an aversive state in which an animal is unable to adapt completely to stressors and the resulting stress and shows maladaptive behaviors' (ILAR 1992).

A variety of factors have the potential to cause significant stress, fear, and distress in laboratory animals. At least in the EU, a sizeable minority of research cats, dogs, and ferrets and most primates originate from non-European countries (Chapter 3). For wild-sourced primates, or those sourced from breeding centres, capture and transportation through a series of stages – some of which are transcontinental – to destination laboratories results in significant stress, fear, and disruption of social networks. Injury, stress-induced *immunocompromisation*, and disease may also result, especially when primates are captured from the wild.

Laboratory housing and environments also have the potential to cause stress and fear, as do both routine and more invasive laboratory procedures. A sizeable minority of all procedures are markedly invasive, and the majority of all procedures do not utilise anaesthetics of any kind. The level of analgesic use is uncertain (Chapter 3).

Animal sourcing

Some primates, in particular, are captured from the wild and transported to breeding centres or laboratories in distant countries (the US, European nations, or elsewhere). Wild capture is usually a highly stressful process, with significant potential for injury. Conditions in holding

29

centres, particularly in the developing countries from which many of these wild-caught primates are sourced, are often suboptimal with respect to safeguarding animal welfare and preventing disease progression and transmission in these stressed, and consequently immunocompromised, animals. Families and other social networks are usually fractured, substantially contributing to psychological distress, which also affects primates sourced from breeding centres. Transport chains from source to destination countries are frequently very lengthy, and trucks or aircraft may subject these animals to potentially alarming visual stimuli, as well as extremes of noise, temperature and pressure changes, physical vibrations, and unexpected movements. Such stressful stimuli are also experienced by cats, dogs, ferrets, and other transported species. It is reasonable to assume that most of these stressors are likely to cause significant stress and fear.

Housing and environment

Rodents and rabbits represent around 80 per cent of all laboratory animals used in the EU (Chapter 3), and proportions are similar internationally. Almost all are mice or rats. Studies have indicated that these species value opportunities to take cover, build nests, explore, socialise with compatible *conspecifics* (members of the same species), and exercise control over subsequent social relationships (Balcombe 2006). However, to standardise experimental conditions, and to facilitate access for experimental procedures and cage-cleaning, laboratory animals are typically kept in small cages, with a minimum of environmental enrichment materials. More enriched cages do exist, but their use is not widespread.

To minimise the potential for cross-infection, animals undergoing procedures are sometimes isolated. Areas in which animals are housed or subjected to procedures usually lack windows and, therefore, natural lighting. Furthermore, animal laboratories are surprisingly noisy places, with noise levels often reaching 90–100 dB during working hours. To put these figures in perspective, a subway train generates around 95 dB (Baldwin & Bekoff 2007). For the overwhelming majority of laboratory animals, conditions such as these persist for a large proportion of their lives.

A range of deleterious psychological and behavioural effects have been linked to such environments. Rodents, for example, engage in excessive grooming, aggression, and *stereotypical behaviours* such as repeated jumping or bar biting. Such repetitive, apparently purposeless behaviours are believed to indicate psychological distress which is both profound and chronic. Empathetic capacities in rodents and other species also allow stress experienced by one animal to affect others – effectively creating a positive feedback loop when animals are closely confined, as occurs in standardised laboratory cages. Fighting may also increase significantly. However, because rodents are largely nocturnal, researchers often remain unaware of the true prevalence of stereotypical behaviour or increased aggression (Baldwin & Bekoff 2007). Many significant physiological distortions have been linked to such conditions. Examples include increases in heart rate, blood pressure, and levels of serum *corticosterone* (a stress hormone), decreased immunocompetence, hypersensitivity to toxins, and increases in a range of pathologies, including tumours, gastric ulcerations, intestinal inflammation, and capillary damage (Van Loo *et al.* 2001, Baldwin & Bekoff 2007).

Even adverse neuroanatomical effects have been documented, including decreased brain cell genesis in mice (Ehninger & Kempermann 2003), and in rats effects on cognition and memory (Paylor *et al.* 1992, Woodcock & Richardson 2000), learning (Faverjon *et al.* 2002), and recovery from brain injury (Passineau *et al.* 2001).

Surprisingly, perhaps, even genetic studies may be significantly affected. Cudilo and colleagues (2007), for example, found that aorta defects associated with the absence of a certain gene almost vanished when affected mice were housed in larger, slightly enriched cages.

Unfortunately, such deleterious effects are often not prevented by the addition of limited enrichment devices. This is hardly surprising when one considers the natural environments and lives of the species involved, and, in many cases, their correspondingly high levels of motivation to seek concealment from potential predators, explore their environment, forage, build nests, defend territories, and engage in cooperative social or reproductive behaviour with compatible conspecifics. Considerably more substantial alterations to cage sizes and contents would be necessary to effectively ameliorate many of the adverse effects experienced by laboratory animals, and the associated distortions of experimental results (Balcombe 2006).

Routine procedures

Nearly all laboratory animals experience routine procedures such as human contact and handling, cage movement and cleaning, injections, *venipuncture* (blood sampling), and weighing. In toxicity studies the use of orogastric *gavaging* – the insertion of a tube from the mouth into the oesophagus for the direct administration of test compounds – is relatively common. Accordingly, the stress-inducing potential of such procedures is of great significance. Yet more markedly invasive procedures garner much greater attention, although these affect a far smaller pool of animals. The potential for routine laboratory procedures to result in stress and fear is often overlooked.

Many of the most common laboratory species suffer significant stress, fear, and possibly distress as a predictable consequence of three routine laboratory procedures that have been documented in at least 80 studies. These procedures are non-invasive handling, venipuncture, and orogastric gavaging. For each of these procedures, effects such as stress and fear have been indicated in many species by the distortion of a range of associated physiological parameters such as serum levels of corticosterone, glucose, growth hormone, and *prolactin*, as well as heart rate, blood pressure, and behaviour. The distortions studied were rapid, pronounced, and statistically significant, although handling elicited variable immune system responses. Unsurprisingly, given these immunological disturbances, increased predisposition to a range of pathologies was also apparent. Physical presence when these procedures were conducted on other animals also significantly elevated physiological parameters indicative of stress in monkeys, mice, and rats. The fact that the animals studied did not readily habituate to these procedures is highly significant when one considers the welfare and ethical problems created over the longer term (Balcombe *et al.* 2004).

Invasive procedures

Nearly all laboratory animals are higher vertebrates, such as mammals, birds, reptiles, and amphibians (Chapter 3). All such animals possess well-developed neuroanatomical mechanisms – including sensory cells and nerve endings (*nociceptors*) and peripheral and central neuroanatomical architecture – that confer the ability to detect and perceive as painful a variety of noxious stimuli, including mechanical, chemical, electrical, and thermal insults. Such mechanisms evolved partly to encourage avoidance of natural agents capable of causing tissue damage.

These same mechanisms may result in pain perception when these animals are exposed to invasive procedures, noxious stimuli, or tissue damage secondary to artificially inflicted diseases or toxic agents, in laboratories. Most, if not all, procedures result in at least mild physical discomfort (e.g. during restraint and venipuncture), and some result in marked discomfort or pain. Around 29–44 per cent of Canadian procedures, for example, are markedly invasive (Figure 3.5). Although analgesic provision is adequate in some cases, it is less so in others, partly due to concerns – well founded or otherwise – that experimental outcomes may be altered by drug use. Some 59–69 per cent of British procedures, for example, do not utilise any kind of anaesthetic (although analgesic use is unknown) (Figure 3.6).

While anaesthetic and analgesic use undoubtedly alters normal physiology, claims that such distortions are so damaging to the investigation of specific hypotheses being tested that their exclusion is required warrant careful scrutiny. Despite increasing recognition (e.g. Sager 2006, Mayer 2007) that pain relief improves both animal welfare and research quality – via minimisation of pain-related physiological, psychological, and behavioural distortions – pain monitoring and analgesic provision remain less than optimal in many research protocols (Hawkins 2002, Flecknell 2008).

The case of chimpanzees

Chimpanzees and other great apes possess a range of advanced psychological and social characteristics, which may enhance their potential for suffering in laboratory environments.

Emotional capacities

The potential suffering of laboratory chimpanzees is compounded by their relatively advanced emotional capabilities. They appear to be able to experience a range of emotions, similar to those we label happiness and sadness, fear and anxiety, irritation, rage, and despair (e.g. Goodall 1995 & 2003), and they appear able to suffer emotional, as well as physical, pain (Fouts 1995). Psychological stress is likely to result both from aversive experiences directly and from the inability of laboratory chimpanzees to escape them. These considerations are of greatest concern where pain or discomfort is substantial or prolonged.

Psychological abilities

The relatively advanced capacities of chimpanzees to understand and remember that certain people, equipment, or procedures are likely to cause pain and distress, and their ability to anticipate future aversive experiences, are likely to compound the distress such events may cause. Chimpanzees have some capacity to anticipate and understand the intentions and psychological states of others (e.g. Wood *et al.* 2007), and have long memories (e.g. Beran 2004). The psychological abilities of chimpanzees may encompass abstract reasoning (Call 2003), self-awareness (although mirror self-recognition may decline with age; e.g. De Veer *et al.* 2003), and simple problem solving (e.g. Goodall 2003). These relatively advanced abilities most probably evolved to enable chimpanzees to cope with their complex natural environments and social structures (Goodall 1995).

Social characteristics

Chimpanzees are highly social animals, and the disruption of social networks when animals are captured from the wild – as many older research chimpanzees once were – or when they are subjected to confinement or translocation prior to or during biomedical research may add to their suffering. The social relationships of chimpanzees appear to encompass prolonged rearing of offspring, close and affectionate family bonds, friendship, and mourning behaviour following the deaths of companions (Goodall 1995). Anecdotal accounts of consolation of victims of aggression and solicitous treatment of injured individuals suggest that chimpanzees feel empathy (Itakura 1994, Silk *et al.* 2005). Chimpanzees plan for the future, and interact in a variety of cooperative activities, including territorial patrols, coalitionary aggression, cooperative hunting, food sharing, and joint mate guarding (e.g. Silk *et al.* 2005).

Chimpanzees possess well-developed communicative skills. Facial expressions (Parr 2003) and sophisticated vocalisations (e.g. Kojima *et al.* 2003, Izumi & Kojima 2004) convey information, for example, about identity (Kojima *et al.* 2003), emotional states (Parr 2003), and social status (Izumi and Kojima 2004). Chimpanzees kiss, hold hands, pat one another on the back, embrace, tickle, punch, and swagger (Goodall 1995), with gestural dialects varying between communities (McGrew 1994).

Despite the relatively advanced emotional, psychological, and social sophistication of chimpanzees, certain morally relevant dissimilarities with humans do exist. Recent research suggests that human *altruistic behaviour* – that is, a willingness to assist others who are not genetically related, in the absence of any personal gain, and possibly despite personal costs – provides a key example. In contrast, assistance offered by chimpanzees and other NHPs appears mainly to be limited to biologically related or reciprocating individuals, and is rarely extended to unfamiliar individuals (Silk *et al.* 2005, Jensen *et al.* 2007), although such behaviour has been observed in common marmoset monkeys (*Callithrix jacchus*; Burkart *et al.* 2007).

Effects of invasive research

It is reasonable to expect that the relatively advanced emotional, psychological, and social characteristics of chimpanzees enhance their capacity for suffering in laboratory environments and invasive research protocols. In the opinion of some experts, such characteristics render it impossible to provide laboratory environments that satisfactorily meet their minimum physiological and behavioural requirements (Balls 1995, Smith & Boyd 2002).

The costs to chimpanzees enrolled in biomedical studies include involuntary confinement in laboratory settings, social disruption, and participation in potentially harmful research. Recent studies have established beyond any reasonable doubt that the effects of laboratory confinement and procedures, especially long term, can be severe. Many captive great apes, including chimpanzees recently retired from US laboratories (Bradshaw *et al.* 2008), show gross behavioural abnormalities, such as stereotypies, self-mutilation or other self-injurious behaviour, inappropriate aggression, fear, or withdrawal (Brüne *et al.* 2006, Bourgeois *et al.* 2007).

It is increasingly acknowledged that such abnormal behaviours resemble symptoms associated with human psychiatric disorders such as depression, anxiety disorders, eating disorders, and post-traumatic stress disorder, and that pharmacological treatment modalities similar to those applied to human patients may be appropriate, and indeed morally compelled, for severely disturbed animal patients (Brüne *et al.* 2006, Bourgeois *et al.* 2007). Long-term therapeutic combination with positive reinforcement training, environmental enrichment, and social and environmental modification may be necessary in severe cases (Bourgeois *et al.* 2007).

Summary

Almost all animals used in laboratories are higher vertebrates, such as mammals, birds, reptiles, and amphibians, which possess the neuro-anatomical and psychological capacities necessary to experience significant pain, fear, and psychological distress. In particular, great apes such as chimpanzees possess advanced emotional, psychological, and social capacities, which enhance their potential for suffering in laboratory environments.

A wide variety of stressors have the potential to cause significant stress, fear, and sometimes distress in laboratory animals. These may be associated with the capture of wild-sourced species to supply laboratories or breeding centres, transportation (which may be prolonged for some animals), laboratory housing and environments, and both routine and invasive laboratory procedures.

The stress caused by laboratory housing and environments, routine laboratory procedures, and in all likelihood other stressors such as those associated with wild capture, transportation, and invasive procedures may result in profound, statistically significant distortions in a range of physiological parameters, including cardiovascular parameters and serum concentrations of glucose and various hormones. Behaviour may be markedly altered, and behavioural stereotypies and increased aggression may develop over time, as may alterations in certain neuroanatomical parameters and even cognitive capacities (Balcombe *et al.* 2004, Balcombe 2006, Baldwin & Bekoff 2007).

Unsurprisingly, the chronic stress experienced by most laboratory animals can result in immunocompromisation and increased susceptibility to a range of pathologies. In addition to creating significant animal welfare and ethical problems, such conditions and their effects on laboratory animals may also distort a wide range of experimental outcomes, such as those dependent on accurate determination of physiological, behavioural, or cognitive characteristics in animal models.

Part II
Human Benefits

5

Human Clinical Utility of Animal Models

Biomedical research using laboratory animals is highly controversial. Advocates frequently claim such research is vital for preventing, curing, or alleviating human diseases (e.g. Brom 2002, Festing 2004a), that the greatest achievements of medicine have been possible only due to the use of animals (e.g. Pawlik 1998), and that the complexity of humans requires nothing less than the complexity of laboratory animals to serve as an effective model during biomedical investigations (e.g. Kjellmer 2002). They have even claimed that medical progress would be 'severely maimed by prohibition or severe curtailing of animal experiments', and that 'catastrophic consequences would ensue' (Osswald 1992).

However, such claims are hotly contested (e.g. Greek & Greek 2002a), and the right of humans to experiment on animals has also been strongly challenged philosophically (e.g. Singer 1990, La Follette & Shanks 1994). Moreover, a growing body of empirical evidence casts doubt on the scientific utility of animals as experimental models of humans.

Case studies

In the field of pharmaceutical development, case studies exemplifying differing human and animal outcomes – sometimes with severe adverse consequences for human patients – are sufficiently numerous to fill entire book chapters (e.g. Greek & Greek 2000 & 2002b).

A recent notorious example was *TGN1412* (also known as CD28-SuperMAB), a fully *humanised* monoclonal antibody (one developed in a non-human species and protein-engineered to express human-specific characteristics) that was under development for the treatment of inflammatory conditions such as leukaemia and rheumatoid arthritis

(TeGenero 2006a & 2006b). During a 2006 *phase I* (i.e. first in humans) clinical trial in the UK, TGN1412 resulted in severe adverse reactions including organ failure. All six trial volunteers required intensive care, and one suffered permanent damage. Yet preclinical testing on several primate, rabbit, and rodent species had failed to reveal adverse effects (Bhogal & Combes 2006, Coghlan 2006; see also pp. 118–19).

Another recent notorious example was the arthritis drug Vioxx, which appeared to be safe, and even beneficial to the heart, in animal studies. However, Vioxx was withdrawn from the global market in 2004, after causing as many as 140,000 heart attacks and strokes, and over 60,000 deaths, in the US alone (Graham *et al.* 2005).

Since their commercial introduction in the early 1980s, non-steroidal anti-inflammatory drugs (NSAIDs) have also had a problematic clinical history. Although apparently safe in year-long studies in rhesus monkeys, benoxaprofen (Oraflex) produced thousands of serious adverse reactions in humans, including dozens of deaths, within three months of its initial marketing (Dahl & Ward 1982). Fenclofenac (Flenac) revealed no toxicity in 10 animal species, yet produced severe liver toxicity in humans, and was subsequently withdrawn (Gad 1990). Similar fates befell some other NSAIDs, including zomepirac (Zomax; Ross-Degnan *et al.* 1993), bromfenac (Duract; Peters 2005), and phenylbutazone (Butazolidin; Venning 1983), which produced adverse human effects undetected in animal studies.

Numerous other pharmaceuticals have also been marketed after passing limited clinical trials and more rigorous animal testing, only later to be found to cause serious side-effects or death in human patients. Examples are various antibiotics (e.g. chloramphenicol, clindamycin, temafloxacin), antidepressants (e.g. nomifensine), antivirals (e.g. idoxuridine), cardiovascular medications (e.g. amrinone, cerivastatin, mibefradil, ticrynafen), and many others (e.g. Wallenstein & Snyder 1952, Blum *et al.* 1994, Greek & Greek 2000).

Although 92 per cent of new drugs that pass preclinical testing, which routinely includes animal tests, fail to reach the market because of safety or efficacy failures in human clinical trials (FDA 2004), adverse drug reactions detected after drugs have been approved for clinical use remain common. They are, indeed, sufficiently common to have been recorded as the fourth to sixth leading cause of death in US hospitals (based on a 95 per cent *confidence interval* (CI); Lazarou & Pomeranz 1998), a rate considered by these investigators to be 'extremely high'.

There are also cases of safe and efficacious human pharmaceuticals that would not pass rigorous animal testing because of severe or lethal toxicity in some laboratory animal species. Notable examples are penicillin (e.g. Koppanyi & Avery 1966), Paracetamol (acetaminophen; e.g. Villar *et al.* 1998), and Aspirin (acetylsalicylic acid; e.g. Wilson *et al.* 1977). More rigorous animal testing might well have delayed or prevented the use of these highly beneficial drugs in human patients.

The large number of cases of apparent differences between outcomes in laboratory animals and in human patients may result from several factors. True discordance in results may arise from interspecies differences (see Chapter 7). In addition, flaws may occur during the pharmaceutical development and testing process by which the design, conduct, or interpretation of experiments may fail to sufficiently highlight the risks to human patients. Such flaws are more likely in animal studies than in human clinical trials because the experimental quality of the former is usually significantly lower (see Chapter 7, pp. 89–90).

Systematic reviews

The premise that animal models are generally predictive of human outcomes is the basis for their widespread use in human toxicity testing, and in the safety and efficacy testing of putative pharmaceutical agents and other clinical interventions. However, the numerous cases of discordance between animal and human outcomes suggest that this premise may well be incorrect, and that the utility of animal experiments for these purposes may not be assured. On the other hand, only small numbers of experiments are normally reviewed in case studies, and their selection may be subject to bias. To provide more definitive conclusions, *systematic reviews* of the human clinical or toxicological utility of large numbers of animal experiments are necessary. Experiments included in such reviews should be selected without bias, via randomisation or similarly methodical and impartial means.

In support of this concept, Pound and colleagues (2004) commented that clinicians and the public often consider it axiomatic that animal research has contributed to human clinical knowledge, on the basis of anecdotal evidence or unsupported claims. These constitute an inadequate form of evidence, they asserted, for such a controversial area of research, particularly given increasing competition for scarce research resources. Hence, they called for systematic reviews to examine the human clinical utility of animal experiments. They commenced by examining six existing reviews, which did not demonstrate the clinical utility expected of the experiments in question.

Soon afterwards, the UK Nuffield Council on Bioethics stated, 'It would ... be desirable to undertake further systematic reviews and meta-analyses to evaluate more fully the predictability and transferability of animal models.' They called for these reviews to be undertaken by the UK Home Office, in collaboration with the major funders of research, industry associations, and animal protection organisations (Nuffield Council on Bioethics 2005).

Since then, several such reviews and *meta-analyses* have been published by other scientists which collectively provide significant insights into the human clinical and toxicological utility of animal models. Meta-analyses combine and statistically analyse data from multiple experiments. To locate such studies, in 2007 I searched the Scopus biomedical bibliographic databases for systematic reviews of the human clinical or toxicological utility of animal experiments published in the peer-reviewed biomedical literature (Knight 2007a & 2008b). These databases are among the world's most comprehensive. At that time they included over 12,850 academic journals, hundreds of other publications, and a total of 29 million abstracts (Scopus 2006).

To minimise bias, reviews were included only when they had been conducted systematically, as described previously. For example, in some cases all the animal studies contained in large subsets of toxic chemical databases were examined, without discrimination.

I examined only reviews which considered the human toxicological predictivity or utility of animal experiments, or their contributions to the development of diagnostic, therapeutic, or prophylactic interventions with clear potential for combating human diseases or injuries, or their consistency with human clinical outcomes. Reviews which focused only on the contributions of animal experiments to increased understanding of the aetiological, pathogenetic, or other aspects of human diseases, or on the clinical utility of animal experiments in non-human species, were excluded from consideration.

In total, 27 systematic reviews which examined the utility of animal experiments during the development of human clinical interventions (20), or in deriving human toxicity classifications (7), were located. Three different approaches which sought to determine the maximum human clinical utility that can be achieved by animal experiments were of particular interest.

Invasive chimpanzee research

Chimpanzees are the species most closely related to humans, and consequently most likely to be predictive of human outcomes when used in biomedical research. However, their similarities to humans, combined with the particularly high costs of their procurement and maintenance, also raise exceptional animal welfare, ethical, and financial concerns when chimpanzees are confined in research laboratories and subjected to invasive experiments. Consequently, chimpanzee experimentation is the subject of considerable debate in those very few countries where it persists.

Apparently seeking to counter increasing international opinion against invasive chimpanzee experimentation, advocates have recently begun extolling its alleged benefits, calling for its continuation. In a prominent plea in *Nature* for increased funding for such research, several heads of US primate research centres stated that chimpanzee experimentation has been of critical importance during struggles against major human diseases (VandeBerg *et al.* 2005). Similarly, in 2006 British scientists called for the right to conduct such research on chimpanzees, contrary to the existing UK ban, in rare scenarios, such as the investigation of dangerous emerging infectious diseases (Jha 2006).

Rapid international developments in this field, including the development of legislation restricting great ape experimentation, justify a detailed examination of the merits of invasive experiments on chimpanzees. Such a reappraisal is most applicable to the US. Although US animal research is governed by international, federal, and state laws, regulations, rules, guidelines, and standards (Meyers 1983), contrary to the legislation of other key countries, the US *Animal Welfare Act* (1966, with standards for laboratory animals amended in 1985) does not require the use of non-animal alternatives, even when scientifically validated alternatives exist. Unsurprisingly, therefore, US primate use has been estimated to be more than five times the number used in the entire EU (approximately 58,000 *v.* 11,000 annually; Conlee *et al.* 2004).

In any assessment of the merits of invasive chimpanzee experimentation, a necessary first step is to obtain a definitive overview of the disciplines investigated by such research. Accordingly, in 2005 I surveyed three major biomedical bibliographic databases and examined published studies conducted worldwide from 1995 to 2004 (Knight 2007b & 2008c). I sought to assess the value of research on captive chimpanzees, particularly when it was invasive, because such research raises the greatest bioethical and social concerns.

I included studies of captive chimpanzees or their tissues, and excluded studies of free-living populations, veterinary medical case reports of naturally ill chimpanzees (whether or not in captivity), most genome studies, studies of skeletal anatomy (which frequently used museum specimens), and studies of *cell lines*, although I did include cell samples, such as peripheral blood mononuclear cells, obtained from captive chimpanzees.

I located 749 studies of chimpanzees or their tissues that met my inclusion criteria, of which 48.5 per cent (363/749) were biological experiments, and 41.5 per cent (311/749) were virological experiments (Figure 5.1).

Biological studies were conducted in nine broad disciplines (Figure 5.2), of which the most common were cognition/neuroanatomy/neurology (36.6 per cent, 133/363) and behaviour/communication (20.7 per cent, 75/363).

In the 41.5 per cent of all chimpanzee experiments that were virological studies (311/749), 30 viruses were investigated (Figure 5.3), of which the most frequent were hepatitis C virus (HCV) and human immunodeficiency virus (HIV) (both 31.2 per cent, 97/311).

The remaining experiments were therapeutic investigations (3.5 per cent, 26/749) – namely, pharmacological, toxicological, and anaesthesiological investigations, and the testing of surgical techniques or prostheses; investigations of eight parasitic species (3.1 per cent, 23/749) – of which the most frequent were the malaria protozoa *Plasmodium falciparum* and *P. ovale* (26.1 per cent, 6/23), the roundworm *Onchocerca volvulus* (21.7 per cent, 5/23), and the flatworm *Schistosoma mansoni* (17.4 per cent, 4/23); and other diseases and miscellaneous experiments, which jointly constituted 3.5 per cent (26/749) of all chimpanzee experiments.

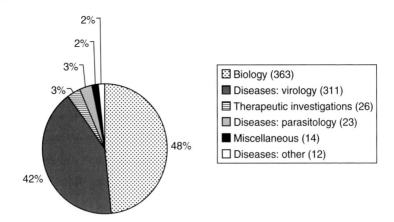

Figure 5.1   Surveyed chimpanzee experiments, 1995–2004 (total 749)
Sources: Knight (2007b & 2008c).

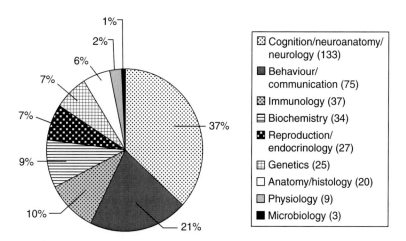

Figure 5.2   Chimpanzee biological experiments, 1995–2004 (363 of 749)
Sources: Knight (2007b & 2008c).

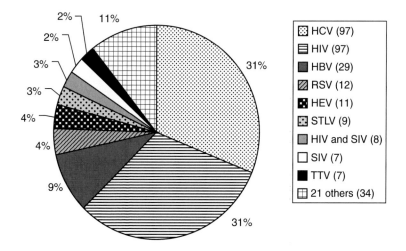

Figure 5.3   Chimpanzee virological experiments, 1995–2004 (311 of 749)

HCV, hepatitis C virus; HIV, human immunodeficiency virus; HBV, hepatitis B virus; RSV, respiratory syncytial virus; HEV, hepatitis E virus; STLV, simian T-cell lymphotropic virus; SIV, simian immunodeficiency virus; TTV, transfusion-transmitted virus.

Twenty-one others – six: foamy virus (human and simian); four: hepatitis A virus; two each: GB virus B, human immunodeficiency virus & herpes virus, influenza virus, parainfluenza virus, noroviruses; one each: bacteriophages, dengue virus, Ebola virus, human cytomegalovirus, hepatitis G virus, human metapneumovirus, human/simian T-cell leukaemia virus, lymphocryptoviruses, papillomaviruses, rhadinovirus (or γ-2-herpesvirus) genogroup 2, rhinovirus, varicella-zoster virus, woolly monkey hepatitis B virus, unspecified.

Sources: Knight (2007b & 2008c).

Advancements in biomedical knowledge?

On the face of it, these studies appear to have contributed to a large array of biomedical disciplines. However, not all knowledge has significant value, nor is worth the animal welfare, financial, or other costs that may be incurred in gaining that knowledge. For a more critical assessment of the utility of invasive chimpanzee research in advancing biomedical knowledge, I randomly selected a statistically significant subset of 95 experiments, and determined the frequency with which they were cited by papers subsequently published and included in these comprehensive bibliographic databases. I found that 49.5 per cent (47/95; 95 per cent CI 39.6–59.4 per cent) were not cited by any subsequent papers (Figure 5.4).

Given that almost all of these chimpanzee experiments would have been approved by at least one institutional ethics committee (institutional animal care and use committee (IACUC) in the US), entrusted with ensuring that these experiments' animal welfare-related, bioethical, and financial costs were reasonably likely to be exceeded by their expected benefits, it is disturbing that half of these randomly selected experiments were not cited by any subsequent papers. The year of publication did not appear to affect this outcome substantially, as citation frequencies were similar across the decade, with more recent papers cited approximately as often as older papers.

Citation frequencies are not, of course, a definitive indication of the benefits, or lack thereof, of scientific research. Uncited studies may also contribute to the advancement of biomedical knowledge, through a variety of mechanisms. However, citation frequencies do generally

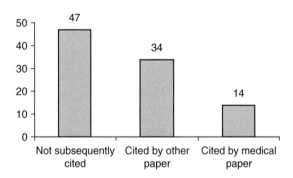

Figure 5.4 Citations of 95 randomly selected published chimpanzee studies
Sources: Knight (2007b & 2008c).

provide a quantifiable and reasonably objective approximation of utility, or lack thereof, and they have been shown to be among the most reliable indicators of research quality (Rochon *et al.* 1994, Callaham *et al.* 2002, Lee *et al.* 2002). Research that makes a significant contribution to a field – such as by confirming or refuting hypotheses – is very likely to be cited by future papers, as is research that produces interesting or controversial outcomes. On the other hand, research that is inconclusive or of little interest or significance is much less likely to be cited.

The disappointing citation rate of these chimpanzee studies is therefore cause for considerable concern. It is extremely unlikely that a study could have provided a significant contribution and remained uncited by any future publication, as was the case for half of these randomly selected chimpanzee studies. In comparison, Callaham and colleagues (2002) found that only 9 per cent of 204 published medical papers they systematically selected and studied were never cited, and that the mean citation rate for these papers overall was two citations annually.

Given that much research of lesser significance is not published, these published chimpanzee experiments can be assumed to be those with the greatest potential for advancing biomedical knowledge. Consequently, these results indicate that the majority of invasive chimpanzee studies generate data of questionable value which make little obvious contribution to the advancement of biomedical knowledge.

Advancements in human healthcare?

Most people would consider that the greatest justifications for invasive chimpanzee research involve attempts to advance human health. As stated, advocates of such research claim it has been critical during our struggles against major human diseases (VandeBerg *et al.* 2005). To assess such claims critically, I determined the frequency with which the statistically significant subset of 95 randomly selected chimpanzee studies had been cited by papers describing diagnostic, therapeutic, or prophylactic methods efficacious in combating human diseases.

Only 14.7 per cent (14/95; 95 per cent CI 8.9–23.4 per cent) of all invasive chimpanzee studies were cited, by a total of 27 papers describing well-developed diagnostic methods (5) or therapeutic and/or prophylactic methods (22) for combating human diseases (Figure 5.4). Diseases and viruses examined included cancer (non-specific), chronic obstructive pulmonary disease, Epstein-Barr virus, hepatitis viruses A through G, hepatocellular carcinoma, HIV, malaria, organ transplant rejection, respiratory syncytial virus, rheumatoid arthritis, rhinovirus colds, systemic lupus erythematosus, and transmissible spongiform encephalopathies.

As stated, citation rates are not a definitive indication of utility or lack thereof. Invasive chimpanzee studies may have contributed to medical advances through various indirect means, such as by stimulating investigation of certain lines of inquiry in humans – although it is unlikely that any medical papers subsequently published would not cite the chimpanzee studies that provided such inspiration. Alternatively, chimpanzee studies may have contributed to investigations of disease *aetiology*, or to papers describing diagnostic, therapeutic, or prophylactic methods for combating human diseases in early stages of development – although potential human benefits, in such cases, remain speculative.

However, it is reasonable to expect that if chimpanzee research had truly been critical during struggles against major human diseases, as claimed by advocates, such chimpanzee studies would, in fact, have been cited by papers describing methods efficacious in combating those diseases. The only alternative is that none of the struggles to which chimpanzee research purportedly made major contributions resulted in effective, published solutions. In fact, 27 papers describing well-developed prophylactic, diagnostic, or therapeutic methods for combating human diseases *did* cite chimpanzee studies. However, detailed examination of these medical papers revealed that *in vitro* studies, human clinical and *epidemiological* studies, molecular assays and methods, and genomic studies contributed most to their development.

For each of these medical papers, the cited chimpanzee studies proved to be of peripheral importance, for one or more of several reasons. Fully 63.0 per cent (17/27) were wide-ranging reviews of 26–300 (median 104) references to which the cited chimpanzee study made a very small contribution. In 12 cases the chimpanzee studies appeared redundant, because humans or human sera were studied concurrently, or because they served only to confirm previous human observations. In seven cases the method explored in the cited chimpanzee study was not developed further, sometimes because later clinical trials in humans failed to demonstrate safety or efficacy, contrary to positive chimpanzee results. In five cases the chimpanzee study examined a disease or method of only peripheral relevance to the medical method described. In three cases the chimpanzee study merely illustrated a historical finding, or was cited only during discussions of previous attempts to combat the disease in question. In another three cases the chimpanzee studies yielded results inconsistent with data from other non-human primate studies, and in two cases they yielded results inconsistent with human data. In two more cases only the human out-comes from studies concurrently described in the cited chimpanzee studies were discussed. In one case the chimpanzee study may have helped establish the need for a new diagnostic method, but did not contribute further to its development. In fact, none of these cited chimpanzee studies demonstrated an essential contribution, or – in most cases – a significant contribution of any kind, to the development of the medical method described.

Since my 2005 survey, several additional studies have been published confirming the lack of utility of chimpanzee experimentation in com-bating major human diseases. In light of calls by some scientists for a return to the use of chimpanzees during AIDS vaccine research and testing, Bailey (2008) compared the published results of AIDS vaccine trials in chimpanzees with those from human clinical trials. Despite the generation of immunogenicity and subsequent protective or thera-peutic responses in chimpanzees following exposure to a wide variety of HIV vaccines, such effects had not been demonstrated by any vac-cine in humans. Bailey concluded that vaccine responses in chimpan-zees and humans are highly discordant, and that a return to the use of chimpanzees in AIDS research and vaccine development would be scientifically unjustifiable.

Similarly, chimpanzee research has not been helpful in the fight against human cancers. Chimpanzee tumours are rare, and profound differences in cell growth, *apoptosis* (normal cell death), carcinogenic-ity and *metastasis* (tumour spread) between chimpanzees and humans severely limit the utility of the former as models for human cancer. Unsurprisingly, therefore, a comprehensive literature search revealed that chimpanzees have rarely been used in cancer research, and that such research has not resulted in effective human clinical interven-tions. The author concluded that cancer research would not suffer if the use of chimpanzees for this purpose were prohibited in the US (Bailey 2009).

Chimpanzees have been widely used in HCV research, but a review of 109 such studies published from 1998 to 2007 revealed that their utility was limited by lack of statistical validity, repeatability, and biological rel-evance of chimpanzees to human HCV. Such studies have not produced a reliable preventive or therapeutic vaccine. In light of the endangered status of chimpanzees, and the particularly high financial and ethi-cal costs of using them in such research, the review's author called for a re-evaluation of HCV research priorities (Bettauer 2010). These con-cerns were echoed by Bailey (2010), who concluded that 'claims of the necessity of chimpanzees in historical and future hepatitis C research are exaggerated and unjustifiable, respectively'.

Experiments expected to yield medical advances

Lindl and colleagues (2005) examined animal experiments conducted at three German universities between 1991 and 1993 that had been approved by animal ethics committees at least partly on the basis of explicit claims by researchers that the experiments might lead to concrete advances towards the cure of human diseases. Experiments were included only where previous studies had shown that related animal research had confirmed the hypotheses of the researchers, and where the experiments had been of sufficient quality to achieve publication in biomedical journals.

For 17 experiments meeting these inclusion criteria, citations were analysed over at least 12 years. Citation frequencies and types of citing papers were recorded: whether they were reviews or animal-based, *in vitro*, or clinical studies. Of the 1,183 citations that were found, only 8.2 per cent (97 citations) were in clinical publications, and only 0.3 per cent (4 citations) demonstrated a direct correlation between the results of animal experiments and human outcomes. However, even in these four cases the hypotheses that had been verified successfully in the animal experiment failed in every respect when applied to humans. None of these 17 experiments led to any new therapies, or had any beneficial clinical impact during the period examined.

As a result, Lindl and colleagues called for serious, rather than cursory, evaluations of the likely benefits of animal experiments by animal ethics committees and related authorities, and for a reversal of the current paradigm in which animal experiments are routinely approved. Instead of experiments being approved because of the possibility that benefits might accrue, Lindl and colleagues suggested that where significant doubt exists, laboratory animals should receive the benefit of that doubt, and that such experiments should not be approved.

Highly cited animal experiments

Hackam and Redelmeier (2006) also used citation analysis, but without geographical limitations. On the assumption that findings from highly cited animal experiments would be most likely to be subsequently tested in clinical trials, they searched for experiments with more than 500 citations that had achieved publication in one of the seven leading scientific journals when ranked by citation impact factor.

Of 76 animal studies located, with a median citation count of 889 (range 639–2,233), only 36.8 per cent (28/76) were replicated in randomised human trials. Disturbingly, 18.4 per cent (14/76) were contradicted by randomised trials, and 44.7 per cent (34/76) had not translated to clinical trials. Ultimately, only 10.5 per cent (8/76) of these medical interventions were subsequently approved for use in patients, and, as noted earlier on page 40, even in these cases human benefit cannot be assumed, because adverse reactions to approved interventions are a leading cause of death (Lazarou & Pomeranz 1998).

A low rate of translation to clinical trials of even these highly cited animal experiments was apparent, despite 1992 being the median publication year, allowing a median of 14 years for potential translation. For studies that did translate to clinical trials, the median time for translation was 7 years (range 1–15 years). The frequency of translation was not affected by the species used, the type of disease or therapy under examination, the journal, year of publication, methodological quality, or even – surprisingly – the citation rate. However, animal studies incorporating dose–response gradients were more likely to be translated to clinical trials (*odds ratio* [OR] 3.3; 95 per cent CI 1.1–10.1).

Although the rate of translation of these animal studies to clinical trials was low, as Hackam and Redelmeier stated, it is higher than that of most published animal experiments, which are considerably less likely to be translated than these highly cited animal studies published in leading journals. Furthermore, the selective focus on positive animal data, while ignoring negative results (*optimism bias*), was one of several factors proposed to have increased the likelihood of translation beyond that scientifically merited. As Hackam (2007) stated, the rigorous meta-analysis of all relevant animal experimental data would probably significantly decrease the translation rate to clinical trials.

In addition, only 48.7 per cent (37/76) of these highly cited animal studies were considered to be of good methodological quality. Despite their publication in leading scientific journals, few included the random allocation of animals to test groups, any adjustment for multiple hypothesis testing, or the *blinded* assessment of outcomes. (The methodological quality of animal experiments is explored further in Chapter 7, pp. 89–90.) Accordingly, Hackam and Redelmeier cautioned patients and physicians about the extrapolation of the findings of even highly cited animal research to cases of human disease.

Stroke and head injury models

Despite the existence of literature on the efficacy of more than 700 drugs in treating experimental models of stroke (artificially induced focal cerebral *ischaemias*), at the time of writing only recombinant tissue plasminogen activator (rt-PA) and Aspirin had convincingly demonstrated efficacy in human clinical trials of treatments for acute ischaemic stroke (NIND & SSG 1995, CAST 1997, ISTCG 1997). Hence, Macleod and colleagues (2005a) stated, 'This failure of putative neuroprotective drugs in clinical trials represents a major challenge to the doctrine that animals provide a scientifically-valid model for human stroke.'

At least 10 published systematic reviews have described the poor human clinical utility of animal experimental models of stroke and head injuries (Horn *et al.* 2001, Jonas *et al.* 2001, Curry 2003, Macleod *et al.* 2005a & 2005b, van der Worp *et al.* 2005, Willmot *et al.* 2005a & 2005b, O'Collins *et al.* 2006, Perel *et al.* 2007).

In some cases, clinical trials proceeded despite equivocal evidence of efficacy in animal studies. For example, Horn and colleagues (2001) systematically reviewed 20 animal studies of the efficacy of nimodipine, of which only 50 per cent showed beneficial effects following treatment. They concluded that 'the results of this review did not show convincing evidence to substantiate the decision to perform trials with nimodipine in large numbers of patients'. These clinical trials also demonstrated equivocal evidence of efficacy, and furthermore proceeded concurrently with the animal studies, despite the fact that the latter are intended to be conducted prior to clinical trials, to facilitate the detection of potential human toxicity.

O'Collins and colleagues (2006) conducted a very large review of 1,026 experimental drugs for acute stroke that had been tested in animal models. They found that the effectiveness in animals of 114 drugs chosen for human clinical use was no greater than that of the remaining 912 drugs not chosen for clinical use, thereby demonstrating that effectiveness in animal models had no measurable effect on whether or not these drugs were selected for human use. Accordingly, O'Collins and colleagues questioned whether the most efficacious drugs are, in fact, being selected for clinical trials, and called for greater rigour in the conduct, reporting, and analysis of animal experiments.

In many cases, animal models did indicate efficacy, but this did not translate to humans. In a few reviews, the authors speculated about the possible causes. For example, Jonas and colleagues (2001) hypothesised that the poor clinical efficacy of neuroprotectants effective in animal models was due to differences in the timing of treatment initiation.

Curry (2003) hypothesised that the human clinical failure of 14 neuroprotective agents successful in animal models was due to the *antagonism* of glutamate – which may be associated with neuroprotection – by drug treatment in clinically normal individuals. He therefore proposed that clinical trials be restricted to real stroke patients, who experience elevated plasma glutamate levels. However, by the time of writing, such speculation had not yet resulted in improvements in the poor clinical record of neuroprotectants successful in animal models.

The utility of the majority of these animal studies also appears to have been impeded by their poor methodological quality. Examples are 20 animal studies of the efficacy of nimodipine (Horn *et al.* 2001); animal studies of the efficacy of melatonin (Macleod *et al.* 2005a); 29 animal studies of the efficacy of FK506 (Macleod *et al.* 2005b); 45 animal studies of five compounds from different classes of alleged neuroprotective agents – clomethiazole, gavestinel, lubeluzole, selfotel, and tirilazad mesylate (van der Worp *et al.* 2005); 25 animal studies of the efficacy of nitric oxide (NO) donors and L-arginine (Willmot *et al.* 2005a); and 73 animal studies of the efficacy of NO synthase inhibitors (Willmot *et al.* 2005b).

The methodological quality of animal studies was typically scored on the basis of characteristics such as appropriate animal models (aged, diabetic, or *hypertensive* animals are considered to more closely model human stroke patients); *power* calculations of sample sizes; random allocation to treatment and control groups; use of a clinically relevant time window for treatment initiation; blinded drug administration; use of anaesthetics without significant intrinsic neuroprotective activity (ketamine, for example, may alter neuroprotective activity); blinded induction of ischaemia (given that the severity of induced *infarcts* may be subtly affected by knowledge of treatment allocation); blinded outcome assessment; assessment of both infarct volume and functional outcome; adequate monitoring of physiological parameters; assessment during both the acute (1–6 days) and chronic (7–30 days) phases; statement of temperature control; compliance with animal welfare regulations; peer-reviewed publication; and conflict of interest statements. Typically, one point was given for the presence of each characteristic. For example, the Stroke Therapy Academic Industry Roundtable recommendations for standards with regard to preclinical and restorative drug development involve an eight-point scale (STAIR 1999, Horn *et al.* 2001).

Median quality scores were 4 out of 10 (13 studies, range 0–6; Macleod *et al.* 2005a; 4 out of 10 (29 studies, range 0–7; Macleod *et al.* 2005b); 3 out of 10 (45 studies; van der Worp *et al.* 2005); and 3 out of

8 (73 studies, range 1–6; Willmot *et al.* 2005b). Common deficiencies included lack of sample size calculations, aged animals or those with appropriate *comorbidities*, randomised treatment allocation, blinded drug administration, blinded induction of ischaemia, blinded outcome assessment, and conflict of interest statements. Some studies also used ketamine anaesthesia, and there was substantial variation in the parameters assessed.

Van der Worp and colleagues (2005), for example, concluded that the collective evidence for neuroprotective efficacy which formed the basis for 21 clinical trials was obtained in animal studies with a methodological quality that could not, in retrospect, justify such a decision. Willmot and colleagues (2005a) also found considerable variations in animal experiment protocols concerning animal species; physiological parameters (such as blood pressure); drug administration (timing, dosage, and route); surgical methodology; and duration of ischaemia. Statistical analysis (Egger's test) also revealed the likely existence of *publication bias* (an increased tendency to publish studies in which a treatment effect is apparent, or a decreased tendency to publish such studies, which may result, for example, from commercial pressures, particularly in the case of patented drugs under development). Macleod and colleagues (2005a) commented, 'These deficiencies apply to most, if not all, of the animal literature.' This is of particular concern because Macleod and colleagues (2005b) reported that efficacy was apparently lower in higher-quality studies, which raised concerns that apparent efficacy may have been artificially elevated by factors such as poor methodological quality and publication bias.

A related review, not limited solely to stroke, exemplified some of these issues. Perel and colleagues (2007) examined the following therapeutic interventions with unambiguous evidence of a treatment effect (benefit or harm) in clinical trials: corticosteroidal treatment for head injury; anti-fibrinolytics for the treatment of haemorrhage; thrombolysis, and also tirilazad, for the treatment of acute ischaemic stroke; antenatal corticosteroids in the prevention of neonatal respiratory distress syndrome; and bisphosphonates in the treatment of osteoporosis. They found that three interventions had similar outcomes in animal models, while three did not, suggesting that the animal studies did not reliably predict the human outcomes. Perel and colleagues also reported that the animal studies varied in methodological quality and sample sizes, that randomisation and blinding were rarely reported, and that publication bias was evident.

Other animal experiments

Of seven systematic reviews on the utility of animal models in other clinical fields identified by my review (Scheld 1987, Lucas *et al.* 2002, Roberts *et al.* 2002, Lee *et al.* 2003, Mapstone *et al.* 2003, Corpet & Pierre 2005, Corry & Kheradmand 2005, Lazzarini *et al.* 2006 – of which Roberts *et al.* 2002 and Mapstone *et al.* 2003 described a single review), in only two cases – one of which was contentious – did the animal models appear to be clearly useful in the development of human clinical interventions, or substantially consistent with human clinical outcomes.

As in the case of stroke, some clinical trials proceeded despite equivocal evidence of efficacy in animal studies. In a systematic review of the effects of low-level laser therapy (LLLT) on wound healing in 36 cell or animal studies, Lucas and colleagues (2002) found that an in-depth analysis of the studies with the highest methodological quality showed no significant pooled treatment effect. Despite this, clinical trials proceeded. Furthermore, almost from the beginning of LLLT investigations, animal experiments and clinical studies occurred simultaneously, rather than sequentially. The human trials also failed to demonstrate significant benefits.

Roberts and colleagues (2002) and Mapstone and colleagues (2003) systematically reviewed a group of 44 randomised, controlled animal studies of the efficacy of fluid resuscitation in bleeding animals. A previous systematic review of clinical trials of fluid resuscitation had found no evidence that the practice improved outcomes, and had even identified the possibility that it might be harmful (Roberts *et al.* 2001). In this later review, the investigators found that fluid resuscitation reduced mortality in animal models of severe haemorrhage, but increased mortality in those with less severe haemorrhage.

After clinical trials in humans failed to provide evidence of benefit, Lee and colleagues (2003) conducted a systematic review and meta-analysis of controlled trials of endothelin receptor blockade in animal models of heart failure. Meta-analysis failed to provide evidence of overall benefit, and indicated increased mortality with early administration.

In their investigation of the contributions of human clinical trial results and analogous experimental studies to asthma research – one of the most common and heavily investigated of modern diseases – Corry and Kheradmand (2005) demonstrated that failure to conduct, and analyse the results of, animal studies before proceeding to clinical trials is not uncommon: 'Research along two fronts, involving experimental models of asthma and human clinical trials, proceeds in parallel, often with investigators unaware of their counterpart's findings.'

The clinical utility of animal models is clearly questionable in cases in which clinical trials proceed concurrently with, or prior to, animal studies, or continue despite equivocal evidence of efficacy in animals. As in the case of stroke, the clinical utility of many of these animal studies also appears to have been limited by their poor methodological quality. Examples are 36 cell or animal studies of the effects of LLLT on wound healing (Lucas *et al.* 2002); 44 studies of the efficacy of fluid resuscitation in bleeding animals (Roberts *et al.* 2002, Mapstone *et al.* 2003); and studies of the efficacy of endothelin receptor blockade in animal models of heart failure (Lee *et al.* 2003). Common flaws included inadequate sample sizes, leaving studies underpowered, and lack of randomisation and blinding.

In some cases, obvious deficiencies in the animal models were identified. In commenting on the clinical relevance of animal models for testing the effects of LLLT on wound healing, Lucas and colleagues (2002) noted that the animal models excluded common problems associated with wound healing in humans, such as ischaemia, infection, and necrotic debris.

Difficulties were also apparent in at least one case in translating animal outcomes to human clinical protocols. Lazzarini and colleagues (2006) reviewed experimental studies of osteomyelitis to ascertain their impact on the *systemic* antibiotic treatment of human osteomyelitis. Although they found that most of the animal models reviewed were reproducible and dependable, they also found that the human predictivity of these studies was unclear, and was possibly undermined by difficulties in establishing the right dose regimen in animals. Although they considered that the use of antibiotic combinations was associated with better outcomes in the majority of animal studies, and that these studies did provide indications of appropriate minimum treatment durations, they concluded that these studies had limited relevance to clinical practice.

In two cases, reviewers reported that animal and human outcomes were substantially consistent, although in one case this conclusion was contentious. While reviewing therapeutic approaches to streptococcal endocarditis, Scheld (1987) reported good overall correlations among results obtained by *in vitro* susceptibility testing (especially killing kinetics in bacterial cultures), in animal experiments, and in clinical trials of different antimicrobial regimens in humans with streptococcal endocarditis.

To investigate the efficacy of rodent models of carcinogenesis in predicting treatment outcomes in humans, Corpet and Pierre (2005) conducted a systematic review and meta-analysis of colon cancer chemoprevention studies involving the use of aspirin, β-carotene, calcium, and wheat bran in rats, mice, and humans. Controlled intervention studies of the recurrence of adenomas in human volunteers were compared with chemoprevention studies of carcinogen-induced tumours in rats and of polyps in GM mice. In the meta-analysis, 6,714 humans, 3,911 rats, and 458 mice were included.

Corpet and Pierre found that comparable results were achieved in rats and humans with aspirin, calcium, β-carotene, and wheat bran. Comparable results were found in mice and humans with aspirin, but discordant results were obtained with calcium and wheat bran (the equivalent β-carotene results were not available). Corpet and Pierre concluded that these results suggest that the use of the rodent models can roughly predict treatment effects in humans, but that the prediction is not accurate for all agents, and that the carcinogen-induced rat model is more predictive than the mouse model used. However, few agents were tested, and two of the three agents tested in mice produced different outcomes in humans, so the conclusion that rodents are predictive of human treatment effects, albeit only roughly, is itself contentious.

Additional citation analyses

A small number of studies not included in my 2007 review have also examined the utility of animal experiments by means of citation analysis. Dagg (1999) found that many animal-based psychological and neurological experiments receive no or few citations following publication. This was also true of animal-based studies in the field of cancer research. On average such studies attracted significantly fewer citations than studies involving humans or tissues derived from humans or very small numbers of animals (Dagg 2000). After examining 594 animal-based studies conducted by investigators affiliated with a major Canadian research hospital, Dagg and Seidle (2004) again demonstrated that animal-based studies are more likely to receive no or few citations than to receive many citations following publication.

Summary

The premise that laboratory animal models are reasonably predictive of human outcomes is the basis for their widespread use in the safety and efficacy testing of drugs and other clinical interventions. However, systematic reviews of the human clinical utility of large numbers of animal experiments selected without bias do not support this assumption. In only 2 of 20 systematic reviews located in a comprehensive search of the biomedical literature did the animal models clearly appear useful in the development of human clinical interventions, or substantially consistent with human clinical outcomes. Furthermore, the conclusions of human predictivity arising from one of these reviews were highly questionable (Knight 2007a & 2008b).

Three reviews, each of which used citation analyses, sought to determine the maximum human clinical utility that might be achieved by animal experiments. One examined published studies of invasive chimpanzee research, because chimpanzees are the species most closely related to humans, and consequently most likely to be generally predictive of human outcomes in biomedical research. However, around half of these published chimpanzee experiments were never cited by any subsequent publication. Given that many experiments remain unpublished, this indicates that the majority of invasive chimpanzee studies generate data of questionable value which make little obvious contribution to the advancement of biomedical knowledge. Only around 15 per cent of all invasive chimpanzee studies were cited by papers describing well-developed diagnostic, therapeutic, and/or prophylactic methods for combating human diseases. However, examination of these medical papers revealed that none of the cited chimpanzee studies demonstrated an essential contribution, or in most cases a significant contribution of any kind, to the development of the clinical method described (Knight 2007b & 2008c).

Another approach examined experiments approved by animal ethics committees at least partly on the basis of explicit claims that the experiments might lead to concrete advances towards the cure of human diseases. However, when the resultant published experiments were examined more than a decade later, less than 10 per cent of a large pool of papers citing the experiments were in clinical publications, and none of these experiments led to any new therapies, or had any beneficial clinical impact, during the period studied (Lindl *et al.* 2005).

The remaining approach examined animal studies with very high citation counts, published in leading scientific journals, because the interventions studied are most likely to be subsequently tested in human clinical trials. Even among these studies, however, nearly half had not been so tested, which usually indicates the emergence of safety or efficacy concerns about the product under development. Only just over a third of these animal studies had translated to human clinical trials and had demonstrated similar human outcomes. Furthermore, despite their publication in leading journals, less than half of these animal studies were of good methodological quality. The poor methodological quality of most animal experiments doubtless contributes to their poor translation rate to human clinical trials, and their frequent discordance with human outcomes.

Such results indicate that the utility of many animal experiments in advancing human healthcare or even biomedical knowledge of significance is poor. This concurs with the findings of other studies which have demonstrated that animal experiments in a variety of fields often receive no or few citations following publication (Dagg 1999 & 2000, Dagg & Seidle 2004), and that their citation rates are lower, on average, than those for studies of humans or tissues (Callaham *et al.* 2002, Dagg & Seidle 2004).

6
Human Toxicological Utility of Animal Models

Due to limited human toxicity data, the identification and regulation of exposure to potential human toxins has traditionally relied heavily on animal studies. However, systematic reviews have indicated that animal studies lack utility for this purpose in the fields of carcinogenicity (at least five reviews: Tomatis & Wilbourn 1993, Haseman 2000, Huff 2002, Ennever & Lave 2003, Knight *et al.* 2006a) and *teratogenicity* (one review: Bailey *et al.* 2005).

The *sensitivity* of a test is its ability to detect compounds possessing the property of interest (*true positives*), and the *specificity* is its ability to detect compounds lacking the property of interest (*true negatives*) – thereby avoiding *false positives*. The sensitivity of animal tests to a range of human toxins highlighted by one review (Olson *et al.* 1998) generally appears to be accompanied by poor specificity, resulting in a high incidence of false positive results.

The importance of carcinogen identification

The risks incurred by reliance on animal test data during human hazard assessment are exemplified by animal carcinogenicity testing. The overwhelming importance of accurate carcinogen identification in comparison to other toxin classes becomes clear when their role in the EU REACH chemical testing programme (see Chapter 2) is examined.

The precise causes of cancer are often difficult to discern owing to the contributions of lifestyle, environmental, occupational, and genetic factors. Nevertheless, some 32,500 cancer deaths, representing around 3.5 per cent of the annual EU total, are considered to arise mainly from occupational exposure to carcinogens. Around 6,500 (20 per cent) of these 32,500 deaths are estimated to arise from exposure to unknown chemical carcinogens (with the remainder associated with known or suspected carcinogens, and thus covered by existing legislation).

In a study conducted for the EC's Environment Directorate General, Postle and colleagues (2003) estimated that by identifying such unknown carcinogens the REACH testing programme might prevent between one-third and two-thirds of these deaths in the EU annually, that is, between 2,200 cancer deaths (0.2 per cent of the EU total) and 4,300 deaths (0.5 per cent).

Compared with most other diseases, the human and economic costs of cancer are very high. Postle and colleagues estimated the cost of a cancer death as between €1.39 million (lower estimate) and €2.14 million (best estimate). These estimates were initially sourced from the UK government, and were based on the amounts surveyed individuals stated they would be willing to invest to reduce their personal risks of transport-related fatalities. In addition, the best estimate included some element of medical costs, and lost production and human costs. These estimates were then adjusted to reflect the age of those at risk of cancer-related death (typically, elderly) and the fact that there is usually a period of ill health prior to a cancer death, which surveyed individuals also stated they would be willing to pay to avoid (a 'cancer premium'). Costs excluded or potentially underestimated included diagnostic and treatment-related medical costs; productivity losses to employers and society; administrative, management, and legal costs incurred by employers; and government expenditure on sick pay and disability benefits.

On the basis of these conservative estimates of the cost of cancer deaths and these lower and upper estimates of the number of cancer deaths likely to be prevented through the REACH system, the cancer-related economic benefits of implementing the system over 30 years were estimated by Postle and colleagues to be between approximately €18 billion and €54 billion (Table 6.1).

Table 6.1 Estimated EU economic benefits over 30 years of reducing annual cancer deaths through implementation of the REACH system (€ million)

	Lower predicted benefit (2,167 deaths prevented annually)	Upper predicted benefit (4,333 deaths prevented annually)
Lower estimate	17,592	35,183
Best estimate	27,083	54,167

Sources: Postle *et al.* (2003) and Knight *et al.* (2006d).

In comparison, the economic benefits of implementing the REACH system for all non-cancer diseases combined were estimated to be between €23 million and €225 million, despite the over-optimistic assumption that the number of these cases related to unknown chemicals would be effectively reduced to zero. These estimates do not represent the total economic value of the REACH system, because the positive effects for wider public health and the environment have not been incorporated. Postle and colleagues focused only on the occupational health benefits for chemical industry workers and downstream users of chemicals. Nevertheless, it seems likely that the accurate identification of unknown carcinogens offers greater benefits for occupational and wider public health than the identification of all other toxin classes combined, when considered from an economic viewpoint alone. The same might reasonably be expected of the human costs. Consequently, the accurate identification of unknown human carcinogens is a top priority for the REACH system.

Animal carcinogenicity testing

Since the first chemical bioassay in 1915, when Yamagiwa and Ichikawa showed that coal tar applied to rabbit ears caused skin carcinomas (Huff 1999), several thousand have been conducted, with the objective of determining the risks posed to humans by the great majority of chemicals for which adequate human exposure data are lacking. However, despite the heavy reliance of governmental regulatory agencies on animal carcinogenicity testing, this remains a controversial area of animal research.

Proponents of the bioassay claim that all known human carcinogens that have been studied in sufficient animal species have produced positive results in one or more species (Wilbourn *et al.* 1986, Tomatis *et al.* 1989, Rall 2000). Critics respond that, if sufficient animal testing is conducted, carcinogenesis will eventually occur in *some* species regardless of cancer risk. A study published in the journal *Mutagenesis* found that of 20 human *non*-carcinogens, 19 were known to be animal carcinogens (Ennever *et al.* 1987). Other investigators have also reported the poor specificity of bioassays for human carcinogens, and have noted the considerable biological and mathematical complexities of attempting to accurately extrapolate animal carcinogenicity data to humans (Monro 1996, Meijers *et al.* 1997, Gold *et al.* 1998).

Other key disadvantages of animal carcinogenicity studies are their protracted time frames and their substantial demands on human, animal, and financial resources. Monro and MacDonald (1998) estimated that typical lifetime rodent carcinogenicity bioassays take at least three years to plan, execute, and interpret. Unsurprisingly, therefore, by 1998 only about 2,000 (2.7 per cent) of the 75,000 industrial chemicals in use and listed in the EPA's Toxic Substances Control Act Inventory had been tested for carcinogenicity (Epstein 1998). The cost of testing these 2.7 per cent of industrial chemicals was millions of animal lives (Monro & MacDonald 1998, Gold *et al.* 1999), millions of hours of work by skilled personnel (Gold *et al.* 1999), and many millions of dollars (Ashby 1996, Greek & Greek 2000).

Ashby (1996) estimated that the bioassay evaluations of 400 chemicals by the US National Toxicology Program (NTP) from the 1970s to the 1990s cost hundreds of millions of dollars. By 2005, the bioassay results of 6,153 experiments on 1,485 chemicals were included in the comprehensive Berkeley-based Carcinogenic Potency Database (Gold *et al.* 2005). Greek & Greek (2000) estimated that the cost of carcinogenicity bioassays exceeds $250 million annually.

Similarly, millions of animal lives have been consumed by animal carcinogenicity studies. Monro and MacDonald (1998) estimated that a single carcinogenicity bioassay may use over 1,200 animals. Furthermore, data from the US (Stephens *et al.* 1998) and Canada (CCAC 1998) indicate that testing procedures such as carcinogenicity studies account for most of the studies that produce the highest levels of pain and distress in laboratory animals. Nor is that pain and distress short term. As exemplified by the US National Cancer Institute (NCI)/NTP protocol, dosing in the standard rodent bioassay begins at 6–8 weeks of age and continues for 90–110 weeks – a period similar to the natural rodent lifespan – after which the survivors are killed and autopsied (Peto *et al.* 1984).

EPA and IARC surveys

The control of human exposure to various potential carcinogens constitutes the most important use of animal carcinogenicity data. Accordingly, I conducted a study to assess the utility of animal carcinogenicity data in providing human hazard assessments for regulatory purposes (Knight *et al.* 2006a). The US federal agency with core responsibility for regulating exposures to potentially dangerous environmental contaminants is the EPA (EPA 2004a), and the chemicals of greatest concern in the US (EPA 2003) are listed in the EPA's *Integrated Risk Information System* (IRIS) chemicals database, along with their toxicity data and resultant human carcinogenicity assessments (EPA 2004b).

I examined the 235 chemicals assigned human carcinogenicity classifications among the 543 catalogued in the EPA's IRIS chemicals database as of 1 January 2004 (EPA 2004c) (Table 6.2). I determined the proportion for which the EPA was able to derive classifications of 'probable human carcinogen' (B2) or 'probable human non-carcinogen' (E) primarily on the basis of animal carcinogenicity data. The relatively few definite classifications of 'human carcinogen' (A) relied primarily on available human exposure data, as did some additional 'probable human carcinogens' (B1). The remaining classifications of 'unclassifiable' (D) or 'possible human carcinogen' (C) were not considered substantially useful for risk assessment or regulatory purposes, and are excluded from the NTP's authoritative annual *Report on Carcinogens* (NTP 2002).

Of the 160 IRIS chemicals assigned human carcinogenicity classifications primarily on the basis of animal data (Table 6.3), the EPA considered the animal data inadequate to support the substantially useful classifications of probable human carcinogen (B2) or probable human non-carcinogen (E) in the majority of cases (93/160, 58.1 per cent; 95 per cent CI 50.4–65.5 per cent).

Table 6.2 All IRIS chemicals assigned human carcinogenicity classifications by the EPA

EPA human carcinogenicity classification	Chemicals	Percentage of total
(A) Human carcinogen (convincing human data)	11	4.7
(B1) Probable human carcinogen (limited human data)	6	2.6
(B2) Probable human carcinogen (sufficient animal or human data)	64	27.2
(C) Possible human carcinogen (animal data inadequate for stronger classification)	40	17.0
(D) Unclassifiable (animal data inadequate for stronger classification)	53	22.6
(D) Unclassifiable (no animal or human data)	58	24.7
(E) Probable human non-carcinogen (sufficient animal data)	3	1.3
Total	235	

Note: Shaded cells are reproduced in Table 6.3.

Sources: EPA (2004c) and Knight *et al.* (2006a).

Table 6.3   160 IRIS chemicals assigned human carcinogenicity classifications primarily on the basis of animal data

EPA human carcinogenicity classification	Chemicals	Percentage of total
(B2) Probable human carcinogen (sufficient animal or human data)	64	40.0
(C) Possible human carcinogen (animal data inadequate for stronger classification)	40	25.0
(D) Unclassifiable (animal data inadequate for stronger classification)	53	33.1
(E) Probable human non-carcinogen (sufficient animal data)	3	1.9
Total	160	

Sources: EPA (2004c) and Knight *et al.* (2006a).

To assess the reliability of EPA carcinogenicity assessments, I then compared them with those of the World Health Organization's (WHO's) International Agency for Research on Cancer (IARC), as published in its *IARC Monographs Programme on the Evaluation of Carcinogenic Risks to Humans*. Compiled by international working groups of scientific experts, the *IARC Monographs* provide critical reviews and evaluations of the evidence relating to the possible carcinogenicity of a wide variety of agents, mixtures, and exposures. They are recognised as authoritative sources of information, and assist governmental agencies in making risk assessments and in formulating decisions concerning any necessary preventive measures. A 1998 users' survey indicated that the *IARC Monographs* were consulted by various agencies in 57 countries. Around 4,000 copies of each volume were printed for distribution to governments, regulatory bodies, and interested scientists (IARC 1999a).

Of the 177 chemicals that had received a human carcinogenicity classification from the EPA on the basis of human or animal data ((A) and (B1) combined with those in Table 6.3), 128 were also assigned human carcinogenicity classifications by the IARC. These 128 were divided into those considered by the EPA to possess at least limited human data (17 chemicals) and those primarily reliant on animal data (111 chemicals) for their human carcinogenicity classifications.

The consistency of classification between the EPA and IARC was examined for these two groups by comparing the carcinogenicity classification proportions within each group using chi square tests, and by comparing the individual classifications of the 111 chemicals primarily reliant on animal carcinogenicity data for their human carcinogenicity classifications.

A *chi square test* provides a statistical calculation of the probability that two data sets, such as the EPA and IARC human carcinogenicity classifications, are samples from the same underlying data population, and that any observed differences are simply due to random sampling variation. Large chi square values (χ^2) reflect increased probabilities that observed differences are due to real differences in underlying data populations.

For those 17 chemicals considered by the EPA to possess at least limited human data, overall EPA classifications were not found to differ significantly from IARC classifications ($\chi^2 = 0.291$, $df = 1$, $p = 0.5896$; Table 6.4). (χ^2 analysis does not allow comparison when one category lacks any data; hence, acrylonitrile, assessed as the only possible human carcinogen by the IARC but as a probable human carcinogen by the EPA, was excluded, yielding a more conservative result.)

However, for the 111 chemicals considered by the EPA to lack even limited human data, but to possess animal data, EPA and IARC classifications were very significantly different overall ($\chi^2 = 215.548$, $df = 2$, $p < 0.0001$; Figure 6.1). (To permit χ^2 analysis, methacrylate, assessed as unclassifiable by the IARC but as the only probable human non-carcinogen by the EPA, was excluded, yielding a more conservative result.)

The data revealed that the EPA was much more likely than the IARC to assign carcinogenicity classifications indicative of greater human hazard. The proportion of chemicals classified by the EPA as probable human carcinogens (60 chemicals), compared with all other categories (51 chemicals), was very significantly different from the proportion predicted by IARC classifications, in which the equivalent numbers of chemicals were 12 and 99 ($\chi^2 = 215.273$, $df = 1$, $p < 0.0001$). Similar

Table 6.4 IARC classifications of EPA chemicals with significant human data (EPA categories A or B1)

Human carcinogenicity classification	EPA	IARC
Human carcinogen (A)	11	12
Probable human carcinogen (B1)	6	4
Possible human carcinogen	0	1
Total	17	17

Sources: IARC (1972–2004), EPA (2004c), and Knight *et al.* (2006a).

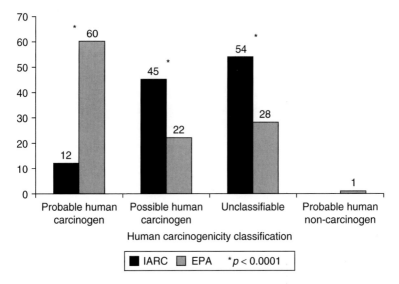

Figure 6.1 EPA and IARC human carcinogenicity classifications of chemicals considered by the EPA to lack human data, but to possess animal data

Source: Knight *et al.* (2006a).

Data source: *IARC Monographs Programme on the Evaluation of Carcinogenic Risks to Humans*, Volumes 1–82 (IARC 1972–2004) and EPA Integrated Risk Information System database (EPA 2004c).

disparities were found for possible human carcinogens (χ^2 = 19.771, df = 1, p < 0.0001) and unclassifiable chemicals (χ^2 = 24.378, df = 1, p < 0.0001).

Comparison of the individual classifications of these 111 chemicals revealed that 67 (60.4 per cent) were assigned an EPA carcinogenicity classification indicative of greater human hazard, 38 (34.2 per cent) were assigned an equivalent classification, and 6 (5.4 per cent) were assigned a classification indicative of lower human hazard than the corresponding IARC classification of the same chemical.

Implications of differing carcinogenicity classifications

On the basis of the EPA figures alone, the predictivity of animal carcinogenicity data for human hazard, and hence their utility in deriving substantially useful human carcinogenicity classifications, is clearly questionable. Among the 160 IRIS chemicals lacking even limited human data but possessing animal data which had received human carcinogenicity assessments, the EPA considered the animal data inadequate to support the substantially useful classifications of probable human carcinogen or non-carcinogen in the majority (93) of cases.

However, IARC assessments of the same chemicals reveal that the human utility of animal carcinogenicity data is probably even lower than indicated by EPA figures alone. EPA and IARC carcinogenicity classifications were similar only for those chemicals with human data. For those lacking human data, the EPA was much likelier than the IARC to assign carcinogenicity classifications indicative of greater human hazard. Of chemicals lacking human data assessed by both agencies, the EPA classified 61 chemicals as probable human carcinogens or non-carcinogens primarily on the basis of their animal data. In contrast, the IARC classified only 12 chemicals similarly, assessing the remainder as unclassifiable or as possible human carcinogens.

The IARC is recognised as one of the most authoritative sources of information on potential human carcinogens (Tomatis & Wilbourn 1993, IARC 1999a), and it is implausible that IARC assessments would often be inaccurate or based on incomplete data. Consequently, the significant differences in human carcinogenicity classifications of identical chemicals between the IARC and the EPA indicate that:

1. in the absence of significant human data, the EPA is over-reliant on animal carcinogenicity data;
2. as a result, the EPA tends to over-predict carcinogenic risk; and,
3. the true predictivity for human carcinogenicity of animal data is even poorer than is indicated by EPA figures alone.

Causes of differing carcinogenicity classifications

Both the EPA and the IARC include a wide range of data in their carcinogenicity assessments. The major categories of evidence considered include human *epidemiological* studies and case reports (or, more rarely, randomised trials), and lifetime exposure studies in test animal species. Conventional chronic rodent carcinogenicity studies use at least 50 animals of each sex per dose group in each of three treatment groups and a concurrent control group, and last for 18–24 months. Assays in genetically engineered rodents may also be considered, particularly with a view to elucidating the chemical or genetic mechanisms of carcinogenesis. Such rodents may have activated *oncogenes* (which cause cancer) introduced (animals with introduced genes are *transgenic*) or tumour suppressor genes deleted or inactivated (*knockout* rodents). Also considered, where available, are supporting data from short-term *genotoxicity* tests, such as standard bacterial and mammalian *in vitro* and *in vivo* tests, and data describing preneoplastic *lesions*, tumour pathology, toxicological effects other than cancer, metabolic and toxicokinetic properties, physicochemical parameters, *structure–activity relationships* (see pp. 102–3), and analogous biologically active agents. On occasion, specific additional studies may be commissioned to fill important knowledge gaps. Both agencies aim to include all pertinent data in deriving a *weight-of-evidence* assessment of human carcinogenic hazard.

However, three key differences between the approaches of the EPA and the IARC may explain their very significantly different human carcinogenicity classifications for identical chemicals:

1. IARC assessments are invariably conducted in depth, whereas EPA assessments may be constrained by available personnel and resources, and may be influenced by human and environmental risks posed by the agent under consideration – which are determined partly by exposure levels. Consequently, EPA staff often conduct screening assessments to decide whether to invest resources in collecting data for a full assessment. Such screening assessments may be based almost entirely on structure–activity relationships and default assumptions, and more detailed assessments may not occur (EPA 2005).
2. While both agencies prefer data from peer-reviewed sources, the IARC appears to be more critical about the standard of the data it accepts. Only occasionally are data accepted from sources other than the peer-reviewed scientific literature, such as government agency reports that have undergone peer review (IARC 2005 & 2006a). This decreases the likelihood that data from animal studies of poor quality (e.g. those with inadequate durations, animal numbers, or survival rates) will be included in IARC assessments (IARC 2006b).
3. As leaders of the US federal agency with core responsibility for protecting Americans from environmental contaminants in the world's most litigious nation, EPA policy-makers understandably err on the side of caution. The impacts on EPA policy are almost inevitable, as illustrated by the EPA *Guidelines for Carcinogen Risk Assessment*, which state, 'The primary goal of EPA actions is protection of human health; accordingly, as an Agency policy, risk assessment procedures, including default options that are used in the absence of scientific data to the contrary, should be health protective. Use of health protective risk assessment procedures as described in these cancer guidelines means that estimates, while uncertain, are more likely to overstate than understate hazard and/or risk' (EPA 2005).

Such policies have affected EPA carcinogenicity assessments for many years. In response to a Congressional directive regarding EPA appropriations for the fiscal year 2000, the EPA undertook an evaluation of data uncertainty and variability in its IRIS assessments. A representative sample of 16 IRIS assessments were subjected to in-depth evaluation by a panel of six independent experts in the field of human health risk assessment. Among other criticisms, they concluded that, despite being advertised as a quantitative, science-based exercise, where uncertainty existed about the data on which decisions were based, the classifications of some chemicals were more reflective of the EPA policy of favouring classifications indicative of greater human risk (Anon. 2000). As noted by the experts, such a policy is consistent with the EPA's mission to protect public health and the environment by means of conservative limits on exposure to possible carcinogens where doubt remains about true human carcinogenic risk.

With respect to the use of animal test data, the EPA *Guidelines* state, 'In the absence of sufficiently scientifically justifiable mode of action information, EPA generally takes public health protective, default positions regarding the interpretation of toxicologic and epidemiologic data ... tumors observed in animals are generally assumed to indicate that an agent may produce tumors in humans' (EPA 2005).

The EPA is strongly defensive of its position: 'The default option is that positive effects in animal cancer studies indicate that the agent under study can have carcinogenic potential in humans ... This option is a public health-protective policy, and it is both appropriate and necessary, given that we do not test for carcinogenicity in humans' (EPA 2005). Despite strong indications from numerous investigations (Shirai *et al.* 1984, Gold *et al.* 1989, Ashby & Purchase 1993, Fung *et al.* 1995, Monro 1996, Meijers *et al.* 1997, Gold *et al.* 1998, Haseman 2000, Rall 2000, Johnson 2001, Ennever & Lave 2003), the EPA seems reluctant to acknowledge the extent of error this may incur, and states, 'The extent to which animal studies may yield false positive indications for humans is a matter of scientific debate' (EPA 2005).

Assessments of other US agencies

EPA carcinogenicity assessments are not necessarily inferior to those of other US regulatory agencies, however. In their survey of 350 representative chemicals, Viscusi & Hakes (1998) found that the human carcinogenicity assessments of other US regulatory authorities, particularly the Food and Drug Administration (FDA) and the Occupational Safety and

Health Administration, were even less based on accurate assessments of carcinogenicity data than those of the EPA.

IARC Monographs surveys

The poor human predictivity of animal carcinogenicity studies was also demonstrated in 1993 by Tomatis and Wilbourn, who surveyed the 780 chemical agents or exposure circumstances evaluated and listed in Volumes 1–55 of the *IARC Monographs* series (IARC 1972–2004). Of these, 502 (64.4 per cent) were classified as having definite or limited evidence of animal carcinogenicity, and 104 (13.3 per cent) as definite or probable human carcinogens. Virtually all of the latter group would, of course, have been members of the former; however, around 398 animal carcinogens were considered not to be definite or probable human carcinogens.

The *positive predictivity* of a test is the proportion of positive test outcomes that are truly positive for the characteristic being tested for (true positives), while the *false positive rate* refers to the proportion that are not (false positives). Hence, on the basis of these IARC figures, the positive predictivity of the bioassay for definite or probable human carcinogens was only around 20.7 per cent (104/502), while the false positive rate was a disturbing 79.3 per cent (398/502).

NTP and other surveys

Surveys by other investigators have also demonstrated the poor human predictivity of animal carcinogenicity data. After examining the studies of 471 substances contained in the NTP carcinogenicity database as of 1 July 1998, Haseman (2000) concluded that although 250 (53.1 per cent) produced carcinogenic effects in at least one sex-species group in which they were tested, the proportion of substances truly posing a significant carcinogenic risk to humans was probably far lower, for reasons such as interspecies differences in mechanisms of carcinogenicity. Similarly, around half of all chemicals, whether natural or synthetic, tested on animals and included in the comprehensive Berkeley-based Carcinogenic Potency Database give positive results (Gold *et al.* 1998).

Rall (2000) estimated that only around 10 per cent of chemicals are truly carcinogenic in humans. Ashby and Purchase (1993) speculated that all chemicals would eventually display some carcinogenic activity if tested in sufficient rodent strains. Even common table salt has been classified as a tumour promoter in rats (Shirai *et al.* 1984).

Fung and colleagues (1995) estimated that if all 75,000 chemicals in use were tested for carcinogenicity via the standard NTP bioassay, significantly less than 50 per cent would prove carcinogenic in animals, and less than 5–10 per cent would warrant further investigation. They suggested that the higher positivity rate recorded was due to chemical selection based on *a priori* suspicion of carcinogenicity. However, examination of the carcinogenicity literature reveals that chemicals are selected for study for many reasons other than *a priori* suspicion, including production volumes, occupational and environmental exposure risks, and investigations of carcinogenesis mechanisms (Gold *et al.* 1989). Despite this, the positivity rate of the carcinogenicity bioassay in the general literature remains around 50 per cent (Gold *et al.* 1998).

Carcinogenicity bioassays fail human validation

Despite heavy reliance on rodent carcinogenicity data during the regulation of human exposures, the conventional rodent bioassay has never been formally validated against human data. On the contrary, validation studies have found this bioassay to be lacking in human specificity (the ability to correctly identify human *non*-carcinogens), resulting in false positive outcomes, or even human sensitivity (the ability to detect human carcinogens at all), depending on the data interpretation method used.

Ennever and Lave (2003) showed that neither of the two commonly used interpretations of rodent carcinogenicity data provides conclusions about human carcinogenicity that are supported by existing human data. If a risk-avoidance interpretation is used, in which any positive result in male or female mice or rats is considered positive, then 9 of the 10 known human carcinogens among the hundreds of chemicals tested by the NTP are positive (Johnson 2001), but so are an implausible 22 per cent of all chemicals tested (Fung *et al.* 1995). If a less risk-sensitive interpretation is used, whereby only chemicals positive in both mice and rats are considered positive, then only three of the six known human carcinogens tested in both species are positive (Johnson 2001). The former interpretation would result in the needless denial of potentially useful chemicals to society, while the latter could result in widespread human exposure to undetected human carcinogens.

Animal teratogenicity testing

In 2005, my colleagues and I published an extensive survey examining the human predictivity of animal teratogenicity testing (Bailey *et al.* 2005). Nearly every putative teratogen tested in more than one species was examined – a total of 1,396 studies. Discordance between species was apparent in just under 30 per cent of these 1,396 reports.

We then analysed in greater detail data for 11 groups of universally acknowledged human teratogens that had each been tested in up to 12 laboratory animal species or species groups – namely, the mouse, rat, rabbit, hamster, primate, dog, cat, pig, ferret, guinea pig, sheep, and cow. Among these species the mean sensitivity for these known human teratogens was only 55.7 per cent. Fully 21.4 per cent of results were equivocal, and false negative outcomes occurred in a disturbing 22.9 per cent of cases. Only around half of these known human teratogens proved to be teratogenic in more than one primate species. We also found that fewer than 1 in 40 of the substances designated as potential teratogens from animal studies were conclusively linked to human birth defects.

We concluded that the poor human predictivity of animal-based teratology warrants the cessation of animal testing, and that resources should be reallocated to the further development and implementation of quicker, cheaper, and more reliable, scientifically validated alternatives, such as the embryonic stem cell test.

Human predictivity of animal toxicity testing

Under the auspices of the International Life Sciences Institute's (ILSI's) Health and Environmental Sciences Institute, Olson and colleagues (1998) sought to determine the extent to which various types of human toxicity evident during clinical trials could be predicted from standard preclinical toxicology studies. On the basis of a multi-company database of 131 pharmaceutical agents with one or more human toxicities identified during clinical trials, they reported a true positive prediction rate of animal tests for human toxicity of 69 per cent, and also that study results from non-rodent (dog and primate) species have good potential to identify human toxicities from many therapeutic classes.

These results concur with those of the other toxicity reviews described. Animal studies are often reasonably sensitive for human toxins. However, their positive predictivity is markedly limited by poor specificity for human toxins, resulting in high false positive rates and a lack of toxicological reliability.

Summary

Because human toxicity data are lacking, the identification and regulation of exposure to potential human toxins has historically and contemporarily relied heavily on animal studies. These are often reasonably sensitive for human toxins; however, poor specificity results in high false positive rates, markedly limiting positive predictivity. Systematic reviews have indicated that animal assays lack reliable predictivity for human toxins in the fields of carcinogenicity (at least five reviews: Tomatis & Wilbourn 1993, Haseman 2000, Huff 2002, Ennever & Lave 2003, Knight *et al.* 2006a) and teratogenicity (one review: Bailey *et al.* 2005).

This was illustrated by my study of the suitability of animal carcinogenicity data in providing human hazard assessments for regulatory purposes (Knight *et al.* 2006a). The environmental contaminants of greatest public health concern in the US are listed in the EPA's IRIS chemicals database (EPA 2004b). Examination of those IRIS chemicals assigned human carcinogenicity classifications primarily on the basis of animal data revealed that the EPA considered those animal data inadequate to derive classifications of probable human carcinogen or noncarcinogen for the majority of these chemicals.

Nevertheless, assessments of the same chemicals by the IARC revealed that the human utility of animal carcinogenicity data is even lower than indicated by EPA figures alone. EPA and IARC carcinogenicity classifications were similar only for those few chemicals with human data. For the remainder, the EPA was much likelier than the IARC to assign carcinogenicity classifications indicating greater human hazard. Likely causes include greater variation in the quality of data accepted by the EPA and in the rigour of its assessments, and public health protective policy erring on the side of caution in what is, after all, the world's most litigious nation. This has clearly resulted in an over-reliance on animal data.

Disturbingly, at least one study found that the human carcinogenicity assessments of other US regulatory authorities were even less accurate than those of the EPA (Viscusi & Hakes 1998).

These findings have profound implications for public health. The accurate identification of unknown carcinogens offers greater benefits for occupational and wider public health than the identification of all other toxin classes combined. Unfortunately, however, the converse is equally true.

7
Factors Limiting the Human Utility of Animal Models

Evaluated overall, the 27 systematic reviews and additional studies examined in Chapters 5 and 6 do not support the widely held assumptions of animal ethics committees and of those advocating animal experimentation that such research is generally beneficial in the development of human therapeutic interventions and the assessment of human toxicity. On the contrary, these studies frequently demonstrate that the utility of animal experiments for these purposes is poor. This appears to result both from limitations of the animal models themselves and from deficiencies in the methodological quality and statistical design of many animal experiments.

Animal model limitations

The scientific limitations incurred through modelling humans by animals in fundamental or clinically applied research and toxicity testing are considerable, wide-ranging, and increasingly recognised. These may include differences between species and sexes, with subsequent effects on *toxicokinetics* and *pharmacokinetics* (PK: the study of bodily distribution, particularly *ADME*: absorption, distribution, *metabolism*, and excretion) or *pharmacodynamics* (PD: the study of mechanisms of action, and drug effects). Additional frequent limitations include loss of biological variability or predictivity resulting from the use of single strains, young animals, restriction to single sexes, and inadequate group sizes; lack of comorbidities or other human risk factors; stress-related physiological or immunological distortions; and the use of unrealistic doses and exposure durations (Hartung 2008a & 2008b, Matthews 2008).

Interspecies variations in P450-dependent monooxygenases, for example, are well established (Guengerich 2006). These constitute the major family of *xenobiotic* metabolising enzymes – enzymes catalysing the oxidation (i.e. the metabolism) of foreign compounds, such as drugs or toxins. Their major purpose is the generation of non-toxic blood-soluble *metabolites* suitable for *renal* (i.e. via the kidneys) or other excretion. Interspecies differences in metabolic pathways, rates, and products may decrease efficacy or produce toxicity, and are a key cause of high clinical trial failure rates during pharmaceutical development (DiMasi *et al.* 2003). In fact, only 8 per cent of all drugs progressing to human trials after demonstration of safety in animal studies gain FDA licensing approval (Pippin 2008).

Chimpanzees: our closest models

Systematic reviews (Knight 2007b & 2008c) indicate that chimpanzees lack utility as experimental models for studying human diseases. On the face of it, this seems counter-intuitive, given the genetic similarities between chimpanzees and humans. Our two species shared a common ancestor just 5–7 million years ago (Kumar *et al.* 2005) – a very short period in *phylogenetic* terms.

A 2005 draft of the chimpanzee genome confirmed it to be 98.77 per cent identical to the mean human genome in terms of base pairs (CSAC 2005). When considering only the most *functional DNA* – that is, bases that cannot be altered without a consequent change in the amino acid coded for by the gene, as distinct from bases that may be altered without such changes, or so-called *junk DNA* outside coding regions – Wildman and colleagues (2003) found a 99.4 per cent correlation between chimpanzees and humans. However, insertions, deletions, and consequent misalignments raise the total estimated difference to 4–5 per cent (Britten 2002, Varki & Altheide 2005).

While a minority of these genetic differences lie within the structural genes which are responsible for all protein production other than regulatory factors, most are now known to lie within the regulatory regions of our DNA. By controlling the activities of structural genes, regulatory genes can exert an 'avalanche' effect on hundreds of other genes. Consequently, a small difference may have profound effects (Bailey 2005). Striking differences have been found in the levels of gene expression between chimpanzees and humans in the brain and liver, for example (Ruvolo 2004). Although chimpanzees and humans differ in only 4–5 per cent of their DNA, this is sufficient to result in a difference of around 80 per cent in protein expression (Glazko *et al.* 2005), yielding marked *phenotypic* differences between the species.

In addition, systemic responses to disease agents and test pharmaceuticals in laboratory chimpanzees may be distorted by endocrinological, immunological, and neurological abnormalities that can result from a variety of experienced stressors. Although chimpanzees may be housed in social groups between studies, with access to enlarged, environmentally enriched enclosures, during study participation laboratory housing may be small, barren, and standardised, and chimpanzees may experience isolation, trauma, chronic boredom, and a variety of stressful laboratory procedures (Fouts 1995).

The substantial differences in protein expression between chimpanzees and humans, and the further distortions of normal physiology that may result from stressful laboratory environments and procedures, confer differences in the susceptibility to, and aetiology and progression of, various diseases; in the absorption, tissue distribution, metabolism, and excretion of pharmaceutical agents; and in the toxicity and efficacy of pharmaceuticals. Such lack of *fidelity*, or accurate reproduction of key human characteristics and responses, is the most likely cause of the demonstrable lack of utility of chimpanzee models during the development of methods efficacious in combating human diseases.

Other laboratory animal species are much less similar to humans, both genetically and phenotypically, and are therefore even less likely to accurately model the progression of human diseases, or human responses to chemicals and test pharmaceuticals, in the great majority of cases.

Toxicity testing

Rodents are by far the most common laboratory animal species used in toxicity studies. Several factors contribute to the inability of rodent bioassays to predict human toxicity reliably. The stresses incurred during handling, restraint, other routine laboratory procedures, and particularly the stressful routes of dose administration common to toxicity tests, alter immune status and disease predisposition in ways which are very difficult to predict accurately. This may distort the progression of diseases and responses to chemicals and test pharmaceuticals (Balcombe *et al.* 2004, Knight *et al.* 2006b).

In addition, animals have a broad range of physiological defences against general toxic insults, such as epithelial shedding and *inducible enzymes*, which commonly prove effective at environmentally relevant doses, but which may be overwhelmed at the high doses often applied in routine toxicity testing (Gold *et al.* 1998). Lower doses, greater intervals between exposures, shorter total periods of exposure, and intermittent feeding, which represent more realistic simulations of the environmental exposure of humans to most potential toxins, might not result in toxic changes at all (Knight *et al.* 2006b).

Finally, differences in rates of absorption and transport mechanisms between test routes of administration and other important human routes of exposure, and the considerable variability of organ systems in response to toxic insults, between and within species, strains, and genders, render profoundly difficult any attempt to predict human hazard accurately on the basis of animal toxicity data (Knight *et al.* 2006b).

These factors are well illustrated by the animal carcinogenicity studies described in Chapter 6. One hundred and sixty environmental contaminants of significant public health concern were listed in the EPA's IRIS toxicity database, and received human carcinogenicity assessments primarily on the basis of animal test data because they lacked significant human exposure data (Table 6.3, p. 65). However, the EPA considered these animal data inadequate to support the substantially useful classifications of 'probable human carcinogen' (B2) or 'probable human non-carcinogen' (E) in the majority (93/160; 58.1 per cent) of cases (Knight *et al.* 2006a).

Species used

The species used were not provided for 2 of these 160 chemicals, both of which were considered unclassifiable with respect to their human carcinogenicity. The other 158 chemicals were each studied in up to seven different species (in the case of *N*-nitrosodiethylamine), with the most common numbers of species used per chemical being two (67.7 per cent), one (14.6 per cent, many, but not all, of which resulted in a human carcinogenicity assessment of 'unclassifiable'), and three (11.4 per cent) (Table 7.1).

At least 10 different species were used for these 158 chemicals, namely, chickens, dogs, guinea pigs, hamsters, mice, mink, primates (one macaque, three unspecified 'monkey' species, and one unspecified 'primate' species), rabbits, rats, and trout. The three species most commonly used were mice (92.4 per cent), rats (86.7 per cent), and hamsters (14.6 per cent) (Figure 7.1).

Discordance among rodents

As in animal studies generally, rodents represented the overwhelming majority of animals used. Intuitively, one might expect mice and rats

Table 7.1 Animal species used for 158 assessed EPA chemicals lacking significant human data

Species used	Chemicals	Percentage of total
1	23	14.6
2	107	67.7
3	18	11.4
4	6	3.8
5	0	0.0
6	3	1.9
7	1	0.6
Total	158	

Sources: EPA (2004c) and Knight *et al.* (2006b).

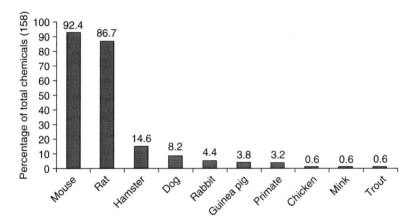

Figure 7.1 Animal species used for 158 assessed EPA chemicals lacking significant human data

Sources: EPA (2004c) and Knight *et al.* (2006b).

to display similar carcinogen susceptibilities. However, while for some chemicals they do, for others they do not. For example, Hengstler and colleagues (1999) described the striking variation in aflatoxin B1 carcinogenesis susceptibility between rats and mice, with mice resistant to dietary levels three orders of magnitude higher than those found to be carcinogenic in rats. The remarkable resistance to aflatoxin B1 of mice results from an efficient conjugation with glutathione, catalysed by glutathione S-transferase mYc – a biochemical mechanism apparently lacking in the rat.

In 2001, Gottman and colleagues (2001) compared 121 replicated rodent carcinogenicity bioassays conducted both within the NCI/NTP and elsewhere, and published and recorded in the comprehensive Berkeley-based Carcinogenic Potency Database. They found a concordance of only 57 per cent overall for rodent carcinogenicity classifications between the NCI/NTP and studies from the general biomedical literature. Disturbingly, this value did not improve substantially when additional biological information (species, strain, sex, target organs) was considered. In addition, Fung and colleagues (1995), Johnson (1999), and Ettlin and Prentice (2002) have demonstrated that bioassays involving strains other than those used by the NTP give discordant results.

Haseman (2000) examined the chronic rodent studies of 385 substances recorded in the NCI/NTP database. After equivocal results were eliminated, 207 chemicals were found to be carcinogenic in at least one sex–species group, yet only 56 of these (27.1 per cent) were carcinogenic in both male and female mice and rats. Male rats appeared to be more susceptible to tumorigenesis than females, but the situation was reversed in mice. Di Carlo (1984) similarly found that of 61 chemicals causing cancer in mice or rats, only 13 (21.3 per cent) were carcinogenic in both male and female mice and rats. He concluded that 'it is painfully clear that carcinogenesis in the mouse cannot now be predicted from positive data obtained from the rat and vice versa'.

Discordance between rodents and primates

The high carcinogenesis predisposition of rodents when compared with primates (Gold *et al.* 1998, Goodman 2001) further complicates the extrapolation of results to humans. Detailed analyses of 25 rodent carcinogens tested in monkeys for up to 32 years by the NCI revealed that half were not monkey carcinogens. In most cases this was contrary to strong evidence of carcinogenicity in rodents and, in some cases, humans (Gold *et al.* 1999). Another 26-year NCI study of model rodent carcinogens found that only urethane was carcinogenic in monkeys (Schoeffner & Thorgeirsson 2000).

Numerous important differences between rodents and humans affect carcinogenesis predisposition. Relevant differences between rats and humans include mean lifespan (2.5 versus 70 years), food consumption (50 versus 10 g/kg/day), basal metabolic rate (109 versus 26 kcal/kg/day), anatomical differences (the forestomach, Zymbal's gland, Harderian gland, preputial gland and clitoral gland exist only in the rat), stomach pH (4–5 versus 1–2), and, very significantly, DNA excision repair rates (low versus high) (Monro & Mordenti 1995).

Species differences in absorption, distribution, metabolism, and elimination pathways or rates can all influence chemical toxicity, including carcinogenicity. Since many carcinogens must be metabolised to reactive electrophiles to produce their carcinogenic effects, species differences in pathways or rates can affect activity levels of carcinogenic metabolites (Dybing & Huitfeldt 1992). Examples of rodent carcinogens considered unclassifiable as to their human carcinogenicity by the IARC because their mechanisms of carcinogenesis are absent in humans include D-limonene (rodent mechanism – renal tubular α2u-globulin-mediated nephrotoxicity), saccharin and its salts, and melamine (rodent mechanism – urolithiasis, partially resulting from rodent urinary composition) (EPA 1991, Hard & Whysner 1994, IARC 1999b).

Indeed, it is remarkable that mice can develop very malignant tumours with multiple genetic alterations within 6–18 months, whereas aggressive tumours in humans or other primates may take many years to reach an equivalently life-threatening stage (Balmain & Harris 2000). Some 50 per cent of all chemicals tested for carcinogenicity in mice or rats are positive in at least one experiment, with carcinogenesis predisposition even higher in some commonly used strains (Gold *et al.* 1998, Gold & Zeiger 2000). Holliday (1996) suggested that the high rodent carcinogenesis predisposition in comparison to humans might be due to less efficient DNA repair, poorer control of genetic stability, and/or altered control of genetic expression. The high doses applied in rodent bioassays also increase apparent carcinogenicity. Metabolic pathways of activation and detoxification may become saturated at high tissue concentrations, resulting in differences in target tissue doses, altered tissue responses, and variations in organs affected (Dybing & Huitfeldt 1992).

Of all the species used in testing, primates are the most biologically similar to humans. However, constraints on their availability, along with high procurement, housing, and maintenance costs and their lengthy lifespans, serve to minimise their use (Schoeffner & Thorgeirsson 2000). In contrast, the widespread availability, low costs, and short lifespans of rodents, which facilitate lifetime exposure and developmental deformity studies, result in the predominance of their use in carcinogenicity bioassays. However, the largely logistical benefits of rodents do not qualify them scientifically as the test species of choice. In fact, the profound discordance of bioassay results between rodents and human beings seriously impedes efforts to derive accurate human carcinogenicity assessments from rodent data.

Routes of administration

Routes of administration were not provided for 4 of the 160 IRIS chemicals examined. Up to 10 routes of administration (in the case of benzo[a] pyrene) were used for each of the other 156 chemicals, with the most common numbers of routes used per chemical being 1 (43.6 per cent), 2 (21.2 per cent), and 3 (19.2 per cent) (Table 7.2).

Twelve non-oral routes of administration and a variety of oral routes, not always specified, were used: *dermal*, inhalation, intramuscular, *intraperitoneal*, intrapleural, intrarenal, intratesticular, intravenous, oral (food), oral (gavage), oral (water), oral (other, e.g. capsule, toothpaste additive), oral (unspecified), subcutaneous, surgical implantation, transplacental, and vaginal painting (Figure 7.2). Those routes most commonly used were food (49.4 per cent), gavage (33.3 per cent), and dermal administration (26.3 per cent). Other routes of major interest were drinking water (21.1 per cent) and inhalation (17.9 per cent).

Route-to-route extrapolation

Judgements frequently need to be made about the carcinogenicity of a chemical via a route of exposure different from that studied. For example, exposures of interest may sometimes occur through the inhalation of a chemical tested primarily through feeding studies (EPA 1999). Given that only 17.9 per cent of the chemicals of interest were tested via inhalation, in contrast to the percentages tested via food (49.4 per cent),

Table 7.2 Number of animal routes of administration tested for 156 assessed EPA chemicals lacking significant human data

Routes of administration	Chemicals	Percentage of total
1	68	43.6
2	33	21.2
3	30	19.2
4	16	10.3
5	4	2.6
6	4	2.6
10	1	0.6
Total	156	

Sources: EPA (2004c) and Knight *et al.* (2006b).

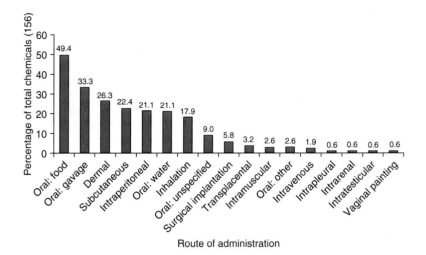

Figure 7.2 Animal routes of administration tested for 156 assessed EPA chemicals lacking significant human data
Sources: (EPA 2004c) and Knight *et al.* (2006b).

gavaging (33.3 per cent) or drinking water (21.1 per cent), such dilemmas hardly appear unlikely.

Quantitative extrapolation between routes of exposure is frequently problematic. Differences in rates of absorption and transport mechanisms between routes (e.g. oral, inhalation, and dermal) can be great. There is no generally applicable method for accounting for these differences in uptake processes, although confidence is strengthened when any tumour effects are observed at a site distant from the point of entry (EPA 1999).

The effects of stress

As described in Chapter 4, studies of mice, rats, hamsters, monkeys, dogs, rabbits, birds, and even bats have shown that common non- or minimally invasive procedures, such as handling or gavaging cause significant increases in stress indicators, including concentrations of glucose, corticosterone (a stress hormone), growth hormone, noradrenaline, prolactin, thyroid-stimulating hormone, and triiodothyronine (Balcombe *et al.* 2004). Other blood measures, including packed cell volume, haemoglobin, and plasma protein, also rise significantly (Gärtner *et al.* 1980). These stress-related responses generally occur with every exposure to such a stressor, and laboratory animals do not readily habituate to these stressors (Balcombe *et al.* 2004).

Stress-related responses are particularly important in long-term carcinogenicity studies, in light of these studies' frequent reliance on stressful routes of administration. Gavaging was used for 33.3 per cent of the 156 applicable EPA chemicals under consideration, and dermal administration (requiring handling and restraint) for 26.3 per cent (Figure 7.2). Other routes of administration normally requiring handling and restraint included intramuscular, intraperitoneal, intrapleural, intrarenal, intratesticular, intravenous, oral (other than food or water, for example, via capsule or toothpaste additive), subcutaneous, surgical implantation, and vaginal painting.

The stress-mediated hormonal changes and disruption of normal hormonal regulation that occur in response to such stressful stimuli predispose animals to immunosuppression and increased susceptibility to virtually all pathologies, including *neoplasia*. Moynihan and colleagues (1990) documented immunosuppression following handling, as evidenced by decreased immunoglobulin G levels in mice. Brenner and colleagues (1990) documented more pulmonary metastases in 11 similarly handled female mice than in non-handled controls following intravenous challenge with tumour cells.

Tumour size can also be affected unpredictably by stressors, complicating assessments. Fifteen female mice handled daily (picked up by the tail and held gently in the palm for two minutes) on days 1–5 of tumour growth had increases in tumour size when compared with non-handled controls, while 13 mice similarly handled on days 1–7 did not (Aarstad & Seljelid 1992).

Intra- and inter-laboratory variations in environmental conditions are likely to result in variations in hormonal disruptions and consequent immunosuppression. For example, animals are positioned in varying proximity to stressful stimuli such as doors, telephones, air-conditioning ducts, and light bulbs. For some years NTP rodent bioassay protocols used different shelves in the animal house for different treatment groups, potentially skewing the results (Young 1989).

To standardise and eliminate such variations in environmental conditions would be a very large-scale and potentially expensive undertaking. Furthermore, it would not eliminate the variable but substantial stresses caused by handling and restraint, and the stressful routes of administration endemic to carcinogenicity bioassays, or their inevitable effects on hormonal regulation, immune status, predisposition to carcinogenesis, and bioassay results.

Organs affected

Of the 160 IRIS chemicals examined, those 3 considered probably not carcinogenic to humans (Table 6.3: category E, p. 65) were not, of course, known to produce significantly neoplastic lesions. For the 53 chemicals with animal data considered unclassifiable (Table 6.3: D), it was frequently difficult to establish whether or not significant treatment-related results occurred. However, for the remaining 104 chemicals considered probable or possible human carcinogens (Table 6.3: B2 or C), up to 43 organs or organ systems were found to exhibit neoplastic lesions (Figure 7.3). However, given that only selected organ systems were examined by some investigators, and that the locations of several gastrointestinal and respiratory tract neoplasms were unspecified, the true prevalence of neoplastic lesions may have been even higher.

Up to 11 organ systems (in the case of 1,2-dibromoethane and N-nitrosodiethylamine) were recorded as exhibiting neoplastic lesions for each chemical (Table 7.3), although most commonly 1 (30.8 per cent), 2 (20.2 per cent), or 3 (16.3 per cent) organ systems were affected. For some chemicals, unclear reporting may have resulted in duplication of results under multiple categories.

The organ systems most commonly affected were the liver (66.3 per cent), the lung (31.7 per cent), and the kidney, skin, and stomach (all 17.3 per cent). Differentiation between primary and metastatic tumours was often uncertain; hence all tumours were included, even when (infrequently) identified as metastatic.

Interspecies variation

The wide variation in organ systems affected may have been exacerbated by the considerable variability of organ systems and species in response to many carcinogenic insults. Comparisons between mice, rats, hamsters, and humans, for example, reveal that carcinogens are carcinogenic at the same site in any two of these species no more than 50 per cent of the time (Gold *et al.* 1998). Patterns of tumour incidence, whether spontaneous or in response to carcinogenic insults, differ dramatically between rodents and humans, and even between different rodent strains. High incidences of rare human tumours, such as those of the liver, pituitary, and testis, occur spontaneously in various strains of mice or rats, whereas common human tumours, such as those of the prostate and colorectum, occur rarely, if ever, in rodents (Monro 1993a & 1993b).

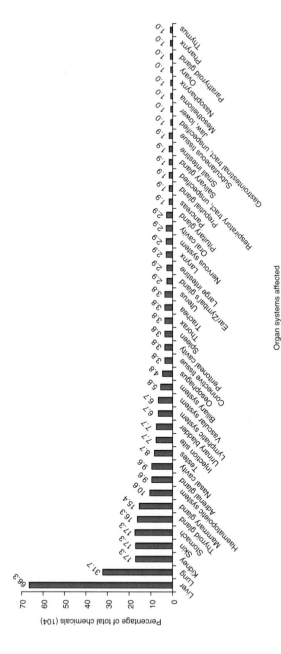

Figure 7.3 The 43 animal organ systems affected by 104 chemicals assessed by the EPA to be probable (B2) or possible (C) human carcinogens

Sources: EPA (2004c) and Knight *et al.* (2006b).

Table 7.3 Number of animal organ systems affected by 104 chemicals assessed by the EPA as probable (B2) or possible (C) human carcinogens

Organ systems affected	Chemicals	Percentage of total
1	32	30.8
2	21	20.2
3	17	16.3
4	9	8.7
5	5	4.8
6	9	8.7
7	5	4.8
8	2	1.9
9	1	1.0
10	1	1.0
11	2	1.9
Total	104	

Sources: EPA (2004c) and Knight *et al.* (2006b).

Dose-related toxicity

In an attempt to interpret differing carcinogenic incidences across a large number of organ systems, investigators commonly assume that human carcinogenic risk is proportional to the number of organ systems affected, and that it increases when multiple sex–species groups are affected or when fatalities result. However, these assumptions were undermined by Lois and colleagues (1991) and Meijers and colleagues (1997). Lois and colleagues analysed around 4,000 chronic carcinogenicity bioassays on 1,050 chemicals, and found that carcinogens affecting multiple sex–species groups or causing fatalities are likely to be the same chemicals that affect multiple organ systems. Meijers and colleagues found that neoplastic lesions in multiple organ systems are more likely to be indicative of dose-related toxicity than true carcinogenicity.

Despite critics such as Monro and Davies (1993), who point out that, since human carcinogens are among the most potent of rodent carcinogens, it should be possible to detect them by using relatively low doses, carcinogenicity bioassays typically rely on *maximum tolerated doses* (MTDs), as indicated by increasing toxicity-related effects, in order to maximise their sensitivity to carcinogens. However, prolonged exposure to high chemical doses can result in chronic irritation, cell death, and consequent cellular proliferation (*mitogenesis*). Sodium saccharin and related sodium and potassium salts, for example, produce urinary calculi, and consequently cellular proliferation, regenerative hyperplasia, and neoplasia, only at high doses (Cohen 1995, Cohen & Lawson 1995, Gold *et al.* 1998).

As stated, animals possess a broad range of physiological defences against toxic insults, including epithelial shedding and inducible enzymes, which commonly prove effective at environmentally relevant doses, but which may be overwhelmed at higher doses (Gold *et al.* 1998). In combination with insufficient rest intervals between doses for the effective operation of DNA and tissue repair mechanisms, the outcome can be increased predisposition to mutagenesis and carcinogenesis. Lower doses, greater intervals between exposures, or shorter total periods of exposure, some or all of which represent a more realistic model of true environmental exposures for most potential carcinogens, might not result in carcinogenesis at all.

Calorie-induced mitogenesis

Reviews of both the experimental and epidemiological literature show a high correlation between increased cell division and carcinogenesis. *Ad libitum* (unrestricted) feeding, as occurs in many carcinogenicity studies, can unnaturally elevate cell division (Gold *et al.* 1998), thereby increasing carcinogenesis predisposition.

Including indices of cell division for both test and control groups would facilitate the identification of groups with elevated cell division rates due to *ad libitum* feeding, thereby allowing increased carcinogenesis predisposition to be accounted for in the interpretation of results. If necessary, feeding protocols could also be altered. However, indices of cell division are not normally determined or considered (Gold *et al.* 1998).

False positive results

It is consequently unsurprising that, for rodent carcinogens tested at the MTD and half the MTD, around half the organs in which tumours arise are affected only at the MTD (Gold *et al.* 1998). Clearly, chronic high-dose carcinogenicity bioassays, in which physiological defences are overwhelmed and *ad libitum* feeding results in the unnatural elevation of cell division rates, are inherently predisposed towards false positive results.

Challenges for human extrapolation

The interpretation of results is also complicated by the following confounding factors: (1) discordance of bioassay results between rodent species, strains, and sexes, and, further, between rodents and humans;

(2) variable, yet substantial, stresses caused by handling and restraint and the stressful routes of administration common to carcinogenicity bioassays, with consequent effects on hormonal regulation, immune status, and carcinogenesis predisposition; and (3) differences in transport mechanisms and rates of absorption between test routes of administration and other important human routes of exposure.

In combination, such factors result in the poor specificity, and hence poor positive predictivity, of animal tests for human carcinogens, as revealed by various investigators (IARC 1972–2004, Ashby and Purchase 1993, Gold *et al.* 1998, Haseman 2000, Rall 2000, Knight *et al.* 2006a). They render profoundly difficult any attempts to extrapolate human carcinogenic hazards accurately from animal data.

The EPA (2006) acknowledged some of the difficulties on its IRIS website:

> In general IRIS values cannot be validly used to accurately predict the incidence of human disease or the type of effects that chemical exposures have on humans. This is due to the numerous uncertainties involved in risk assessment, including those associated with extrapolations from animal data to humans and from high experimental doses to lower environmental exposures. The organs affected and the type of adverse effect resulting from chemical exposure may differ between study animals and humans.

Methodological quality

Among those examined in Chapters 5 and 6, at least 11 systematic reviews (Horn *et al.* 2001, Lucas *et al.* 2002, Roberts *et al.* 2002, Lee *et al.* 2003, Mapstone *et al.* 2003, Macleod *et al.* 2005a & 2005b, van der Worp *et al.* 2005, Willmot *et al.* 2005a & 2005b, Hackam and Redelmeier 2006, Perel *et al.* 2007 – of which Roberts *et al.* 2002 and Mapstone *et al.* 2003 described a single review) demonstrated the poor methodological quality of many animal experiments. None of the 27 reviews noted good methodological quality among a majority of the animal experiments studied. While the omission of experimental details due to journal word-length constraints may artificially lower apparent quality, the prevalence of such deficiencies exceeds what might reasonably be expected, and is, accordingly, grounds for considerable concern.

Common deficiencies included lack of sample size calculations, sufficient sample sizes, randomised allocation to treatment groups, blinded administration of test compounds and assessment of outcomes, and conflict of interest statements. In the experimental studies of stroke reviewed in Chapter 5, some used anaesthetics that may have altered study outcomes, and many failed to ensure the induction of ischaemia was blinded with respect to treatment or control group allocation. These deficiencies limited the clinical utility of these studies in various significant ways. For example, it is well established that studies lacking randomisation or blinding often over-estimate the magnitude of treatment effects (Poignet *et al.* 1992, Aronowski *et al.* 1996, Marshall *et al.* 2000). Bebarta and colleagues (2003) described the impacts of insufficient randomisation or blinding on estimates of the significance of treatment effects in 389 animal and 2,203 cell line studies. They found that studies lacking randomisation or blinding, but not both, were more likely to report a treatment response than studies that used these measures (OR 3.4, 95 per cent CI 1.7–6.9; and OR 3.2, 95 per cent CI 1.3–7.7, respectively), and that studies lacking both randomisation and blinding were even more likely to report a treatment response (OR 5.2, 95 per cent CI 2.0–13.5).

Statistical design

According to Balls and colleagues (2004), 'surveys of published papers, as well as more anecdotal information, suggest that more than half of the published papers in biomedical research have statistical mistakes, many seem to use excessive numbers of animals, and a proportion are poorly designed'. Festing (2004b) agreed: 'Surveys of published papers show that there are many errors, both in the design of the experiments and in the statistical analysis of the resulting data. This must result in a waste of animals and scientific resources, and it is surely unethical.'

In the determination of appropriate sample sizes, a range of factors should be considered (Chapter 9, pp. 125–6). As stated by the UK's Medical Research Council (1993), 'The number of animals used ... must be the minimum sufficient to create adequate statistical power to answer the question posed.' However, De Boo and Hendriksen (2005) observed that animal numbers are often influenced by scientifically irrelevant factors such as availability or cost. This may lead to insufficient sample sizes, leaving many studies underpowered and thereby limiting the statistical validity of study conclusions. Animal lives and resources are then wasted if experiments subsequently require repetition.

Raising standards: evidence-based research

Evidence-based medicine (EBM) bases clinical decisions on methodologically sound, prospective, randomised, blinded, controlled clinical trials. The gold standard for EBM is the large prospective epidemiological study or the meta-analysis of randomised and blinded, controlled clinical trials (Evidence-Based Medicine Working Group 1992).

The application to animal experiments of the EBM standards currently applied to human clinical trials would make the results more robust and would increase their applicability (Watters & Goodman 1999, Moher *et al.* 2001, Arlt & Heuwieser 2005, Schulz 2005, Perel *et al.* 2007). Traditional toxicity testing and assessment would be replaced by *evidence-based toxicology* (Hoffmann & Hartung 2006, Neugebauer 2009). However, mechanisms would be needed to ensure compliance with such standards. Compliance could, for example, be made a prerequisite for research funding, ethics committee approval, and the publication of results. These measures would require the education and cooperation of funding agencies, ethics committees, and journal editors.

The Medical Research Council requires researchers who are planning clinical trials to reference systematic reviews of previous related work before they are permitted to proceed (Pound *et al.* 2004). A similar requirement to reference, or where necessary conduct, systematic reviews of relevant animal studies prior to the commencement of further animal studies would encourage a more complete and impartial assessment of existing evidence (Pound *et al.* 2004, Knight 2008d). This would facilitate the elimination of unnecessary experimental duplication, or the appropriate modification of experimental protocols.

To aid realistic assessments of compounds, mechanisms are also needed to encourage the reporting of negative results. Negative results of preclinical studies are much more likely to remain unpublished than negative results from clinical trials (Brown *et al.* 1995a). In a systematic review of studies of the efficacy of nicotinamide in combating experimentally induced stroke, comparisons published only in abstract form gave a significantly lower estimate of effect size than those published in full – thereby demonstrating publication bias (Oktem *et al.* 2000). Van der Worp and colleagues (2005) commented on the pressure to obtain and publish positive results: 'It is therefore conceivable that the career of a preclinical investigator is more dependent on obtaining positive results, than that of a clinical trialist.'

Summary

Many systematic reviews have demonstrated that animal models are not generally beneficial in the development of human therapeutic interventions or the assessment of human toxicity (Chapters 5–6).

Causative factors include limitations of the animal models used and the poor methodological quality and statistical design of many animal experiments.

Animals frequently fail to accurately mirror human responses with sufficient accuracy (or *fidelity*). Interspecies differences may exist in absorption, distribution, metabolism, and elimination pathways or rates, resulting in differing toxico- or pharmacokinetics. Differences may also occur in toxico- or pharmacodynamics, and all of these may contribute to differences in the organ systems affected, and in the nature and magnitude of those effects.

Biological variability and predictivity are frequently compromised by restriction to single strains, young animals, and single sexes which lack concurrent human risk factors such as comorbidities.

Important human routes of exposure may vary from those tested in animals, requiring extrapolation between routes as well as species. In addition, toxicity tests such as carcinogenicity bioassays typically rely on MTDs and chronic dosing, which may overwhelm physiological defences effective at environmentally relevant doses, resulting in false positive outcomes.

Stresses incurred during handling, restraint, and other routine laboratory procedures, and particularly the stressful routes of dose administration common to toxicity tests, as well as stresses resulting from laboratory environments can alter physiological, hormonal, and immune status in ways which may be difficult to predict. This may alter the progression of diseases and distort responses to chemicals and test pharmaceuticals.

Finally, many animal studies suffer from serious methodological flaws, such as lack of sample size calculations, sufficient sample sizes, randomised allocation to treatment groups, blinded administration of test compounds and assessment of outcomes, and conflict of interest statements, which limit the reliability of any results derived.

Strategies designed to increase the full and impartial examination of existing data before conducting animal studies, to improve methodological quality, and to decrease bias during the publication of results would minimise the consumption of financial and other resources and animal lives in studies of questionable merit and quality, and would increase the robustness, and potentially the utility, of subsequent results. However, the poor human clinical or toxicological utility of many animal experiments is unlikely to result solely from methodological flaws or publication bias. As stated by Perel and colleagues (2007), the failure of animal models to adequately represent human disease may be another fundamental cause, which, in contrast, could be technically and theoretically impossible to correct.

The genetic modification of animal models through the addition of foreign genes, or the inactivation or deletion of genes, is often intended to make them more closely model humans. However, as well as being technically difficult to achieve, such modification may not permit clear conclusions, due to a large number of factors, including those reflecting the intrinsic complexity of living organisms, such as the variable redundancy of some metabolic pathways between species (Houdebine 2007). Furthermore, the animal welfare burdens incurred during the creation and use of GM animals are particularly high (Thon *et al.* 2002, Sauer *et al.* 2006).

Part III
Alternative Strategies

8
Non-Animal Research and Testing Methodologies

Scientific interest in alternatives

Scientific interest in non-animal alternatives has been reflected by their ongoing development and formal validation, and by the establishment of several international centres and university departments dedicated to this work (Leist *et al.* 2008a). In 1981 a Center for Alternatives to Animal Testing was established at the John Hopkins University Bloomberg School of Public Health, to support the creation, development, validation, and use of alternatives to animals in research, product safety testing, and education. In 2004 the British government established the National Centre for the Replacement, Reduction and Refinement of Animals in Research, whose ultimate aim is the *replacement* of all animal experiments. Similar centres exist in Germany, Austria, the Netherlands, and Japan.

Between 1998 and 2010, 30 distinct tests or categories of test method that could replace, reduce, or refine laboratory animal use had been assessed and declared by the European Centre for the Validation of Alternative Methods (ECVAM) – an EC organisation – to be scientifically valid. Twenty-four had achieved regulatory acceptance (ECVAM 2010b).

In 2008 a partnership was announced between the National Institutes of Health (NIH) Chemical Genomics Center of the National Human Genome Research Institute, the National Institute of Environmental Health Sciences, and the EPA (e.g. Leist *et al.* 2008b). The partnership aims to meet the future toxicity testing needs of the EPA and the NTP by implementing recommendations proposed in the NTP's *Roadmap for the Future* (NTP 2004) and the report of the National Research Council's Committee on Toxicity Testing and Assessment of Environmental Agents,

Toxicity Testing in the 21st Century: A Vision and a Strategy (NRC 2007). These reports propose significantly increased roles for non-animal alternatives such as high-throughput robotic molecular and cellular screening methods. They estimate, for example, that running 1,408 chemicals through 200 protein- and cell-based assays, at 15 different concentrations, would take about two-and-a-half weeks, and cost around USD 1 million. In contrast, a traditional rodent bioassay takes upwards of two years to produce results of demonstrably poor specificity for human carcinogens. The carcinogenicity testing of 2,000 industrial chemicals listed in the EPA Toxic Substances Control Act Inventory cost hundreds of millions of US dollars, and consumed millions of animal lives (Chapter 2, p. 63).

Scientific resistance to alternatives

However, resistance to the use of alternatives remains considerable in some governmental, academic, and commercial sectors. A 2000 survey identified a range of common areas of non-compliance of US researchers with the regulations associated with the *Animal Welfare Act*. The most common was inadequate consideration of alternatives (at 600–800 research facilities), and the fourth most common was unnecessary experimental duplication (at approximately 250 facilities). Others included inadequate justification for animal numbers and alleged uncertainty of research personnel about signs indicative of pain and/ or distress (USDA 2000).

Similarly, chemical companies submitting HPV test plans to the EPA have often failed to follow even the minimal guidance provided by the EPA about alternatives. They have failed to use existing published data individually or in conjunction with other data (in a *weight-of-evidence* approach – which aims to review all available relevant evidence in a structured, systematic, independent, and transparent manner; Balls & Combes 2006). They have failed to avoid duplicative or otherwise unnecessary animal testing, and have proposed irrelevant tests (such as acute fish toxicity tests on water-insoluble chemicals) or tests clearly unnecessary to meet HPV requirements. Opportunities to use replacement assays, such as *in vitro* genotoxicity tests, have been ignored, and companies have failed to take advantage of opportunities to combine protocols, sometimes doubling the number of animals killed. In its responses to such test plan proposals, the EPA has frequently failed to encourage companies to follow basic animal welfare principles (Sandusky *et al.* 2006).

Such attitudes towards biomedical animal use reveal a marked lack of consideration of scientific constraints on the utility of animal models for human applications, of social concerns about their use, and of the unprecedented logistical challenges that reliance on animal models will inevitably raise during high-throughput chemical testing programmes. They also demonstrate marked and widespread deficiencies in awareness of the potential and availability of non-animal methods.

Accordingly, this chapter seeks to illustrate in broad terms the potential offered by non-animal (replacement) methodologies, through examination of selected examples. As stated by Gruber and Hartung (2004), 'Availability and awareness of adequate *in vitro* techniques represent the prerequisites for the use of alternative methods,' although alternatives are not limited to *in vitro* methodologies.

This review does not seek to cover any particular sub-discipline definitively within the sizeable alternatives field, or to review *refinement* or *reduction* strategies, which are described in Chapter 9.

Alternatives to laboratory animal use

It is now considered fundamental to good laboratory animal practice to conduct research in accordance with the *3Rs* famously proposed by Russell and Burch in 1959 (e.g. USDA 2000b). These are the:

1. *replacement* of animal use with non-animal alternatives, wherever possible;
2. *reduction* of animal numbers to the minimum possible; and
3. *refinement* of animal use, to avoid or minimise animal pain, distress, or other adverse effects suffered at any time during the animals' lives, and to enhance well-being (Buchanan-Smith *et al.* 2005).

Additional Rs have occasionally been proposed by others. *Re-use* or *recycling* of animals aims to reduce total numbers. *Rehabilitation* aims to ensure protection and care of animals after procedures have been terminated (Anon. 1986, Pereira & Tettamanti 2005).

As stated by Russell and Burch, 'Refinement is never enough, and we should always seek further reduction and if possible replacement.... replacement is always a satisfactory answer.' A broad range of investigative tools exist that may potentially replace sentient animal use in biomedical research and toxicity testing.

Sharing and assessment of existing data

The first step in any toxicological evaluation of a compound, or the commencement of any biomedical research project, should be a comprehensive and critical assessment of all existing data in order to determine which, if any, remaining experiments are necessary. To maximise commercial competitiveness, a great deal of existing data remains excluded from the public domain in pharmaceutical and chemical company files. However, such exclusion is markedly contrary to the public interest when toxicity data regarding the components of new pharmaceutical or consumer products are not disclosed, when animal experiments are repeated, and when product development is slowed as a result. Accordingly, legislation should be enacted for mandatory disclosure of toxicity data. Such regulation need not necessarily be devoid of commercial benefit.

A similar view is held by the European authorities, who have made data sharing mandatory under REACH – to reduce both vertebrate testing and the potential costs incurred by industry through duplicative testing. The REACH guidelines aim to share the costs of studies that have already been conducted among parties that may benefit through access to resultant data. Where additional testing is deemed necessary, the sharing mechanism aims both to minimise costs and to ensure such testing is conducted only once (ECHA 2008).

The Vitic Nexus toxicity database, hosted on the Internet by the scientific education charity Lhasa Limited (www.lhasalimited.org), provides access to public and proprietary carcinogenicity, mutagenicity, hepatotoxicity, skin sensitisation, and other toxicity data for a large array of compounds. The database is 'chemically intelligent', enabling searching for and recognition of similarities in chemical structures. It provides an example of how such information might be organised and publicly shared.

Physicochemical evaluation and computerised modelling

The high costs of drug development have created a strong incentive for the creation of tests able to detect deficiencies such as poor absorption, target organ concentration, clearance or efficacy, and toxicity or other adverse effects early in the development process. The benefits of modelling and simulation during drug design and preclinical assessment can potentially be realised through integration of data on known or expected physicochemical properties, pharmacokinetics, and pharmacodynamics (Balls 1994, Lavé *et al.* 2007, Wishart 2007).

Wishart (2007) recently reviewed available tools for modelling and predicting drug absorption, distribution, metabolism and excretion. Such tools include ADME parameter predictors, metabolic fate predictors, metabolic stability predictors, cytochrome P450 (cP450) substrate predictors, and *physiology-based pharmacokinetic* or *biokinetic* (PBPK or PBBK) modelling software.

PK models predict *in vivo* concentrations of test substances or their metabolites and the organ systems affected by them (Combes *et al.* 2007). PBPK models utilise the known anatomy and physiology of the organism (considering properties such as tissue type (e.g. fat versus muscle), tissue volume, blood flow rates, and known active transport or *biotransformation* pathways), as well as the physicochemical properties of the compound, to estimate the concentration–time relationship of compounds within various body compartments and organ systems (Blaauboer *et al.* 2006). Such models can be used to predict optimal doses and dosing schedules, potency, and, potentially, level of variability and uncertainty for drug candidates (Wishart 2007). They may even be used for extrapolation of the biokinetic behaviour of chemicals between different routes of exposure or species (Andersen 1991, Broadhead & Bottrill 1997).

Germani and colleagues (2007) provided an example of a PBPK model for predicting the plasma concentration–time curves expected after intravenous administration of candidate drugs to rodents. The predictions were based on a small number of properties that were either calculated on the basis of the structure of the candidate drug (e.g. octanol:water partition coefficient, ionisation constant(s)) or obtained from the typical high-throughput screens implemented in the early drug discovery phases (fraction unbound in plasma, intrinsic *hepatic* clearance). Plasma concentration–time curves were predicted with good accuracy.

Grouping and read-across

Chemicals with common structural features or metabolic pathways often have similar physicochemical properties, toxicological effects, and environmental fates. Such compounds can therefore be grouped into chemical categories, allowing those characteristics which are unknown for only some members of the group to be predicted (*read across*) from others. Data thereby read across may be qualitative (such as prediction of the presence or absence of specific toxicological effects) or quantitative (as occurs when levels of toxicity are predicted). An important example is the prediction of *no observed adverse effect levels* (NOAELs) – maximal concentrations or doses at which no adverse changes are detectable.

Structure–activity relationships

The relationship between physicochemical structure and biological activity (*structure–activity relationship*, SAR) was recognised as early as 1894 by 1902 Nobel laureate Emil Fischer, who used the lock-and-key analogy to describe the three-dimensional complementarity of drugs with target receptors or enzymes (Vedani *et al.* 2007a). PD modelling uses SARs to predict various biological activities on the basis of molecular substructures or other chemical moieties. *Quantitative structure–activity relationships* (QSARs) are mathematical descriptions of the relationships between the physicochemical properties of molecules and their biological activities (Comber *et al.* 2003). They may be used to predict characteristics at the theoretical, compound design stage (Vedani *et al.* 2007b).

Due to the large amount of data available for these toxicity endpoints, many *in silico* models exist for the prediction of mutagenicity and carcinogenicity. Ashby and Tennant (1991) demonstrated that the presence of aromatic amino and nitro groups, alkylating agents, and certain other chemical groups increased the likelihood of rodent carcinogenesis in a survey of 301 chemicals tested by the NTP. Cronin and colleagues (2003) described a number of electrophilic molecular substructures common to a range of potential multi-species toxicants. Matthews and Contrera (1998) described the evaluation of a QSAR computerised system that demonstrated 97 per cent sensitivity for rodent carcinogens, and 98 per cent specificity for non-carcinogens.

QSARs are often based around models for particular classes of compounds, such as amines and aldehydes (Benigni *et al.* 2003). Vedani and colleagues (2007b), for example, described their VirtualToxLab, which used an advanced QSAR modelling system to predict interactions of drugs and environmental chemicals with receptors mediating endocrinological disruption or other toxicities. The VirtualToxLab offered 'virtual test kits' for eight receptors: oestrogen α and β, androgen, thyroid α and β, glucocorticoid, aryl hydrocarbon, and peroxisome proliferator-activated γ, as well as for the enzyme cP450 3A4. Tests on 798 compounds predicted binding affinities close to empirically predicted values, with only 6 of 188 test compounds removed by more than a factor of 10 from the known binding affinity, and the maximal individual deviation not exceeding a factor of 15.

Such *in silico* systems are ideally suited for predicting receptor-mediated toxicity caused by drugs and environmental chemicals. Running VirtualToxLab using contemporary mid-level computer technology allowed prediction of binding affinities for 150 compounds over a weekend. Vedani and colleagues estimated that 60–70 per cent of the approximately 30,000 chemicals to be re-evaluated in REACH could be readily processed using VirtualToxLab. To improve the applicability and reliability of VirtualToxLab through peer review, they created a free Internet access protocol available to selected laboratories.

Such models generally have a strong mechanistic basis and are often user-friendly and easily understood (Combes *et al.* 2007). They are fast, reproducible, and cheap (Vedani *et al.* 2005). However, their *applicability domains* are inevitably limited (Combes *et al.* 2007). As long as some enzyme and receptor systems mediating toxicity remain unknown, false negative results will inevitably occur (Vedani *et al.* 2005). QSAR applicability domains will expand as mechanisms and, particularly, receptors mediating additional toxic effects become known. Improved QSAR models for skin sensitisation, respiratory irritation, and genotoxicity, for example, are under development (Veith 2006).

Expert systems

Computerised expert systems seek to mimic the judgement of expert toxicologists by using known rules about factors affecting toxicity, in combination with physicochemical or other information about a specific compound. They make predictions about toxicity and related biological outcomes, such as metabolic fate. Gerner and colleagues (2004), for example, described two sets of *structural alerts* (fragments of chemical structure) for the prediction of skin sensitisation which comprise 15 rules for chemical structures deemed to be sensitising by direct action and 3 rules for substructures that act indirectly – that is, requiring chemical or biochemical transformation. Pre-validation against sensitising chemicals taken from the Allergenliste held by the German Federal Institute for Risk Assessment yielded some reasonable predictivity rates: positive predictivity, 88 per cent; false positive rate, 1 per cent; specificity, 99 per cent; and negative predictivity, 74 per cent; although the false negative rate was 80 per cent and the sensitivity was 20 per cent. The authors proposed that these structural alerts should be subjected to formal validation, with a view to incorporating them into regulatory guidelines thereafter.

Expert systems such as MultiCASE, PASS, DEREK, OASIS/TIMES, and ToxScope aim to predict the mutagenicity of compounds or their major metabolites (Combes *et al.* 2007). Some systems aim to predict carcinogenicity or other toxic endpoints. For example, the EPA has developed Oncologic, an expert system for the prediction of the carcinogenicity of fibres, metals, polymers, and more than 48 classes of organic chemicals, which is freely available via the Internet (EPA 2010b). In comparison to conventional rodent bioassays, such computerised systems are rapid and inexpensive.

Minimally sentient organisms

The use of minimally sentient animals from lower phylogenetic orders, or early developmental vertebral stages, as well as microorganisms, higher plants, and non-living physical and chemical systems, can increase compliance with animal use legislation, which frequently regulates only live vertebrate use.

Cosson (2007), for example, proposed the development of protozoal – rather than rodent – assays for bacterial infection, on the basis that pathogenic bacteria such as *Pseudomonas aeruginosa* often use the same mechanisms to defend themselves against phagocytosis by unicellular amoebae and to infect mammalian cells.

The UK *Animals (Scientific Procedures) Act* (1986) (Anon 1986) allows unregulated use prior to half-way through gestation (in the case of mammals), or incubation (birds and reptiles), or the stage when independent feeding occurs (amphibians and fish). For example, early chicken embryos may be used in reproductive toxicity tests without restriction, and parasites may be similarly cultivated in chicken eggs for use in parasitological studies (Balls 1994, Eckert 1997).

However, European *Directive 2010/63/EU on the protection of animals used for scientific purposes* notes scientific evidence that mammalian foetuses in their last trimester of gestation experience increased risks of pain, suffering, and distress, and accordingly regulates such research (EU 2010).

The use of minimally sentient animals may minimise ethical concerns – although not necessarily to the extent first apparent. For example, there is no clear demarcation between the sentience (and consequent ability to suffer in experiments) of some lower vertebrates, such as certain fish species, and some higher invertebrates, such as certain cephalopod molluscs. However, given scientific evidence of the ability of the latter to experience pain, suffering, distress, and lasting harm, *Directive 2010/63/EU* extends protection to cephalopods (EU 2010).

Similarly, the stage of development at which animals pass from a legislatively unprotected to a protected state is rarely based on strong scientific justification. Such examples demonstrate the necessity of making decisions appropriate to each case on the basis of the best available scientific evidence, and also a precautionary principle mandating that where reasonable doubt exists about sentience, animals should be afforded the benefit of that doubt (Balls 1994).

Tissue cultures

Tissue cultures are *in vitro* cultivations of cells, tissues, organs, or embryos. *Primary cell cultures* are isolated directly from animal tissues, often using proteolytic enzymes. Their major advantages include tissue-specific functions and retention of capacity for biotransformation. However, cellular isolation can result in damage to cell membrane integrity, with damage to, or loss of, membrane receptors and cellular products. Fortunately, during the interval necessary to establish monolayer cultures from cell suspensions, such cell damage is often repaired (Broadhead & Bottrill 1997).

Due to the adverse changes that occur in primary cultures maintained for prolonged periods, it may be necessary to isolate new cells for each experiment. *Cell lines* eliminate this need. However, extended periods of culture over many years can result in decreased metabolic capacity and altered cellular function. Widely used animal-derived cell lines, such as Chinese hamster ovary cells and mouse lymphoma L178Y cells, suffer from variation in phenotype, behaviour, and even chromosome numbers between laboratories (Broadhead & Bottrill 1997).

Immortalised cell lines are capable of extended, and often indefinite, growth *in vitro*. They are generated by introducing viral oncogenes such as SV40 large T, polyomavirus large T, and adenovirus EIA into primary cells using calcium phosphate or electroporation treatment. Examples of cells that have been immortalised are rabbit kidney cells, mouse macrophages, rat *hepatocytes* (liver cells), and human lymphocytes and osteoblasts (Broadhead & Bottrill 1997). Immortalisation of cell lines, however, may significantly alter their characteristics and function (Luttun & Verfaillie 2006).

Stem cell (SC) *lines* can be established from mammalian blastocysts (*pluripotent embryonic stem cells* (ESCs); Huggins 2003) or from *multipotent adult progenitor cells* (Luttun & Verfaillie 2006). Multipotent progenitor cells have the potential to differentiate into cells from multiple, but a limited number of, lineages, whereas a pluripotent cell can create all cell types other than additional embryonic tissue. SCs of either kind can be maintained in an undifferentiated state in the presence of feeder layers and/or purified leukaemia inhibitory factor (LIF) (Brown *et al.*

1995b). On removal of the LIF, SCs differentiate into a variety of cell types (Brown *et al.* 1995b).

The broader the differentiation potential, the more applications can be considered for each stem cell type. The differentiation potential of adult stem cells was long believed to be limited, although recent evidence has challenged this belief. However, SCs derived from human embryos (hESCs) raise considerable ethical and sociological concerns. Nevertheless, as with immortalised cell lines, their self-renewal capacity and potentially unlimited availability make SCs particularly attractive as a research tool (Luttun & Verfaillie 2006).

Much of the considerable interest in SC research centres on their potential to replace lost tissue or functional cells, for example, in degenerative diseases (Luttun & Verfaillie 2006). In a related application, SC research has helped elucidate some of the proliferation and differentiation mechanisms involved in *haematopoiesis* (blood cell production) – and, particularly, the important contribution to this process of several haematopoietic *cytokines* (controlling proteins). These are now used in the treatment of cancer patients undergoing chemotherapy and radiation therapy, which frequently suppress haematopoiesis (Luttun & Verfaillie 2006).

Organ cultures are three-dimensional, and retain some or all of the *histological* features of the *in vivo* equivalent. They may be derived from differentiated SCs. Complex organotypic culture systems, with cofactors and metabolic supplements added, may be used to increase longevity and maintain cellular differentiation (Broadhead & Bottrill 1997).

One important potential application is the creation of artificial liver constructs for biotransformation studies and drug toxicity testing (Luttun & Verfaillie 2006). One of the functionally important structural elements of the liver is the sinusoid, which comprises a fenestrated vascular structure coated with hepatocytes. In the adult liver, sinusoids are radially aligned, running from the periphery of liver lobules to each central vein. Odde and Renn (2000) developed *laser-guided direct writing* (LGDW), enabling the positioning of multiple cell types onto biological matrices with micrometre precision. Cell viability was not significantly impaired (Nahmias *et al.* 2004). These results suggest that LGDW might be used to create a vascular backbone for the *ex vivo* creation of liver constructs. LGDW could also be used to distribute beads coated with different growth factors onto stem cells in order to induce their differentiation in a specific spatial pattern (Luttun & Verfaillie 2006).

Cultures of tissue explants, such as liver slices, can also be used. Their advantages include speed and ease of preparation, the presence of multiple differentiated cell types, and the maintenance of cell–cell and cell–matrix interactions. However, maintenance of viability is difficult, and their use was originally described by Russell and Burch (1959) as *relative replacement*, because animals are killed to provide cells, tissues, or organs, as distinct from the ideal of *absolute replacement*, in which sentient non-human animals are not used at all (Broadhead & Bottrill 1997).

Tissue cultures can play a valuable role in a variety of interesting toxicological investigations. For example, a number of *complementary DNA* (cDNA) segments (from which the non-coding *intron* sequences of the original DNA have been excised) coding for human cP450 enzymes have been cloned, incorporated into *eukaryotic expression vectors*, and introduced into cell lines, which are then used to study the metabolic capacity of individual cP450 enzymes. This may also facilitate identification of compounds that become toxic after biotransformation by these enzymes (Broadhead & Bottrill 1997).

Similarly, cell lines generated to carry *recombinant shuttle vectors* can be used to study eukaryotic mutation, because the vectors can be exposed to mutagenic agents in mammalian cells and then transferred to bacteria – which reproduce much faster – for rapid analysis of the mutations. Such cell lines can be used for short-term *in vitro* mutagenesis studies, DNA repair studies, and studies of the effects of oncogenes, mutator genes, and tumour suppressor genes on mutant frequencies (Broadhead & Bottrill 1997).

Perfused cultures

Tissue cultures may be *static* or *perfused* with culture media via pumps. When exposed to toxins, perfusion cultures may allow continuous real-time monitoring of *cytotoxicity* through detection and measurement of *biomarkers* in the perfusion outflow. This can yield important information about onset time, duration, and scale of toxicity, which is more difficult to obtain from static cultures or *in vivo* assays unless highly specific organic biomarkers indicative of tissue damage are detectable in bodily fluids such as blood or urine. However, the increased complexity of perfused cultures may limit their use to low-throughput analysis (Jennings *et al.* 2004).

Brain tissue cultures

Brain cell cultures provide an example of the therapeutic potential offered by tissue cultures. Bacterial meningitis has a hospital mortality rate of approximately 25 per cent, and up to 50 per cent of survivors suffer neurological sequelae secondary to brain damage. Evidence suggests that lesions result primarily from host immune responses. To date, the *pathogenesis* of brain infections, and the efficacy of potential therapeutic approaches supporting brain repair, have been investigated largely in animal models, many of which experience suffering graded as intermediate to severe. *In vivo* models have been considered necessary because of the complexity of the pathogenesis of meningitis, which involves interactions between the host's inflammatory system, susceptible brain cells, and the bacterial pathogen itself (Leib 2007).

Leib (2007), however, described the development of an *in vitro* model aimed at reproducing important *pathophysiological* processes of damage and tissue regeneration in infectious brain diseases. Objectives include differentiating neuronal stem cells into defined developmental stages susceptible to injury, subsequently challenging them with infectious pathogens and/or inflammatory mediators, and evaluating the integrative and therapeutic potential of neuronal precursors or stem cells as transplants into brain tissue. Brain injury secondary to bacterial meningitis prominently affects the cortex and hippocampus, so Leib developed an organotypic cell culture system capable of facilitating the culturing of hippocampal or cortical brain tissue, while maintaining key elements of natural architecture and natural distribution of cell types.

In vitro assays

In vitro assays using bacterial, yeast, protozoal, mammalian, or human cell cultures exist for a wide range of toxic and other endpoints, including skin corrosion and irritation, phototoxicity, eye corrosion and irritation, skin absorption, skin sensitisation, carcinogenicity, and systemic toxicity (Huggins 2003, Murthy 2007). *In vitro* assays are not without their limitations (Hartung 2007). Nevertheless, some have been scientifically *validated* (demonstrated to be reliable and relevant for the intended purpose; Balls *et al.* 1990), and many others are under rapid development (Huggins 2003). A definitive review of the sizeable and speedily evolving field of *in vitro* assays is well beyond the scope of this chapter; however, examination of selected assays serves to illustrate their broad potential.

Carcinogenicity assays

In vitro assays can contribute information to a weight-of-evidence characterisation of a compound which is sufficient to render the traditional rodent carcinogenicity assay unnecessary. Brusick (1977) found a correlation of approximately 90 per cent between *in vitro* microbial mutagenesis and mammalian carcinogenic properties for a large array of chemicals. Tennant and colleagues (1990) successfully predicted the outcomes for 86 per cent of 44 chemicals undergoing carcinogenicity testing by the NTP by using the *Salmonella* mutagenicity and subacute (90-day) rodent toxicity tests in combination with chemical structural information. The Ames *Salmonella typhimurium* reverse mutation and chromosomal aberration (CA) genotoxicity assays have been accepted by regulatory agencies for many years (Dearfield *et al.* 1991).

Cell transformation assays detect morphological changes that provide the earliest phenotypically identifiable signs of carcinogenicity. These assays were comprehensively reviewed by Combes and colleagues (1999), and the Syrian hamster embryo (SHE) cell transformation assay has since been described as the most predictive short-term assay for rodent carcinogens (Zhang *et al.* 2004). Pienta and colleagues (1977) showed a 91 per cent correlation between the morphological transformation of SHE cells, despite their prior cryopreservation, and the reported carcinogenic activity of numerous carcinogenic and non-carcinogenic chemicals. The particular advantage of the SHE assay in comparison to other *in vitro* assays is its ability to detect some non-genotoxic as well as genotoxic carcinogens (Amacher & Zelljadt 1983, Mauthe *et al.* 2001).

The SHE assay is still being improved. Most of the difficulties encountered in earlier versions have been overcome by culturing SHE cells at pH 6.7 (Kerckaert *et al.* 1998, Zhang *et al.* 2004). In a study of 56 chemicals (30 carcinogens, 18 non-carcinogens, 8 inconclusive), LeBoeuf and colleagues (1996) reported an overall concordance of 85.4 per cent (41/48) between the pH 6.7 SHE cell transformation assay and rodent bioassay results, with a sensitivity of 86.7 per cent (26/30) and a specificity of 83.3 per cent (15/18). Furthermore, the assay exhibited a sensitivity of 77.8 per cent (14/18) for *Salmonella*-negative carcinogens, demonstrating its ability to detect non-mutagenic carcinogens. Both 24-hour and 7-day exposures were used.

Mauthe and colleagues (2001) described the testing of the SHE assay in the Health and Environmental Sciences Institute branch of the ILSI Alternative Carcinogenicity Testing collaboration, which began in 1996. A total of 19 ILSI compounds were tested in the SHE assay, of which 16 were either known rodent carcinogens and/or human carcinogens and 3 were non-carcinogens. The overall concordance between the SHE assay and rodent bioassay results was 89.4 per cent (17/19), whereas concordance with known or predicted human carcinogenicity was a more disappointing 36.8 per cent (7/19). However, Zhang and colleagues (2004) demonstrated the potential for further improvement by demonstrating sensitivity increases by factors of 1.4–2.5 when the cellular incubation time was reduced from the usual 24 to less than 6 hours prior to seeding onto feeder layers.

Test batteries

Batteries of *in vitro* assays may be used to increase the spectrum of toxins detected (*applicability domain*). Kirkland and colleagues (2005) examined the ability of a battery of three of the most commonly used *in vitro* genotoxicity assays – the Ames *Salmonella typhimurium* assay, the mouse lymphoma assay, and the *in vitro* micronucleus or CA tests – to correctly predict the rodent carcinogenicity of 700 known chemicals. While the sensitivity of the battery was high (93 per cent of chemicals with reliable genotoxicity data yielded positive results in at least one test), the specificity was variable. The low specificity of the mammalian cell assays declined further when combined as a battery. However, by adopting *relative predictivity* (RP) *ratios* of true:false results, it was established that positive results in all three tests indicated that rodent carcinogenicity was more than three times as likely as non-carcinogenicity, and also that negative results in all three tests indicated that rodent non-carcinogenicity was more than twice as likely as carcinogenicity. Accordingly, the authors recommended the use of RP ratios to assess the likelihood of carcinogenicity or non-carcinogenicity for chemicals with batteries of positive or negative results.

GreenScreen genotoxicity assays

Lichtenberg-Frate and colleagues (2003) demonstrated the genotoxic and cytotoxic sensitivities of a GM yeast (*Saccharomyces cerevisiae*) assay that used a yeast-optimised version of a green fluorescent protein fused to the RAD54 yeast promoter, which is activated upon DNA damage. The result was green fluorescence in the presence of several genotoxic test compounds. This elegant assay was thereafter known as a 'GreenScreen'.

A similar assay appeared to detect a number of genotoxins and non-genotoxins with a very high level of accuracy in a study of some 75 chemicals (Hastwell *et al.* 2006). Akyüz and Wiesmüller (2003) also described a GreenScreen assay for the detection of genotoxicity, but in this case living human cell cultures were studied. The assay detected the genotoxic effects of ionising radiation, of a compound that tested positive in a traditional Ames bacterial mutagenicity assay, and also of two compounds that were poorly mutagenic in the Ames assay.

As Cahill and colleagues (2004) observed, the spectrum of compounds detected by the GreenScreen is somewhat different from that detected by bacterial genotoxicity assays, and hence the combination of this assay with a high-throughput bacterial screen and an *in silico* SAR screen might provide an effective battery of genotoxicity screening tests for regulatory purposes.

Additional advantages include the possibility of assessing the impact of human metabolites, for example, by the addition of liver extracts, or by using transgenic cell lines with human metabolising enzymes. The potential for automated performance and flow cytometric analysis also makes the human-based assay described by Akyüz and Wiesmüller amenable to high throughput.

Human hepatocyte assays

Human hepatocyte assays play a key role in pharmaceutical development. The liver usually receives the highest *bolus* concentration of an ingested drug, delivered from the intestines via the portal circulation. Furthermore, hepatocytes produce P450-dependent monooxygenases, and hence are the major cells responsible for drug metabolism and the first cells to be affected by any reactive or toxic metabolites produced. Accordingly, human hepatocyte cultures are the *in vitro* system of choice when studying drug-induced processes in humans (Ullrich *et al.* 2007, Li 2008a). They are now routinely used for metabolic profiling (identification of metabolites formed from the parent drug), and the assessment of drug–drug interaction potential and toxicity, in both academic and industrial laboratories. Standardised and validated human hepatocyte culture systems now exist (e.g. Ullrich *et al.* 2007). The development of cryopreservation using protocols and reagents designed to protect membranes from damage associated with cryopreservation and subsequent thawing and the development of protocols and media allowing cultivation of differentiated, functional human hepatocytes have facilitated their widespread use (Ullrich *et al.* 2007, Li 2005 & 2008a).

Assessment of metabolite activity and organ–organ interaction

The inability of most cell cultures to mimic organ–organ interactions occurring *in vivo* represents an important limitation of isolated organ cultures. The importance of considering multiple organ interactions is illustrated by the potential generation of toxic metabolites by the liver which may exert effects in distal organs (Li 2008b). Although such reactive metabolites are believed to constitute a significant proportion of all toxins, for most compounds the degree of toxicity resulting from metabolic activation is unknown (Langsch & Nau 2006).

Hence, models such as the Integrated Discrete Multiple Organ Co-culture (IdMOC) system developed by Li and colleagues (2008b) are particularly significant. Developed for the evaluation of xenobiotic toxicity, the IdMOC system uses a wells-in-a-well structure. Cells from different organs are co-cultured in discrete inner wells, but interconnected by an overlying medium integrated into the outer well. The system models multi-organ human or animal *in vivo* systems, in which organs are physically separated yet interconnected via the systemic circulation. Physical separation of different cell types facilitates evaluation of organ-specific effects, which is extremely difficult when mixed-cell-type co-cultures are used.

Potential applications of the IdMOC system include the evaluation of multiple organ metabolism, as well as organ-specific drug distribution and toxicity. Maximal human relevance is achieved through the use of human xenobiotic metabolising cells, such as hepatocytes, and human target organ cultures. A particularly interesting application is the co-culturing of cancer cells and cells representing major organs for the selection of anticancer agents with minimal organ toxicity (Li 2008b).

Given that pathways of metabolic activation are frequently uncertain – particularly during high-throughput screening – such metabolic activation systems should ideally include all possible relevant hepatic or other enzymes (Langsch & Nau 2006). Such preparations may also be used to identify the metabolic pathways and enzyme systems involved in biotransformation, and the active metabolites produced.

Quality control of biological products

Regulatory safety and potency testing for biological products such as vaccine batches has traditionally relied on animal models. Indeed, the quality assurance (QA) and production of medicines accounts for around 15 per cent of all regulated EU scientific animal use (Leist *et al.* 2008a). Such tests are responsible for a high proportion of experiments resulting in severe, unrelieved pain and suffering (Stephens *et al.* 2002).

However, the *consistency approach* has facilitated reductions in such animal use in the field of vaccine testing. Each vaccine batch is viewed as one of a series produced from the same starting material (seed lot) that shares many characteristics with other batches, rather than as a unique product. This allows for a quality control strategy focusing on production consistency, achieved through compliance with good manufacturing practice (GMP) and QA principles, and assessment at various stages during the production process using non-animal test models such as physicochemical methods and *in vitro* assays. Examples of recent achievements in the quality control of toxoid vaccines include the replacement of challenge procedures by serological methods, and reductions in animal numbers achieved by changing from multi- to single-dose testing (Hendriksen 2006).

Pyrogens (fever-inducing contaminants) can be a major health hazard in *parenteral* drugs when cytokines such as interleukin-1β, interleukin-6, or tumour necrosis factor-α are produced. These cytokines can produce acute adverse reactions, which can range from induction of fever to life-threatening pyrogenic shock with multi-organ failure (as occurred in the case of TGN1412: see p. 118), depending on the concentration of the stimulus and its intrinsic activity (Montag *et al.* 2007). Hence, pyrogen testing may be required prior to batch release. For example, the *European Pharmacopoeia* made such testing of small-volume parenterals obligatory in 2004. This was traditionally achieved using the *in vivo* rabbit pyrogen test (since 1942) or the limulus (horseshoe crab) amoebocyte lysate (LAL) test (since 1976).

However, six variants of human immune cell-based assays have been validated in a collaborative international study. Such human *in vitro* assays can achieve sensitivities and specificities of greater than 90 per cent for human pyrogens, and – in contrast to the LAL test – may detect non-endotoxic pyrogens derived from Gram-positive bacteria or fungi (Schindler *et al.* 2006a & 2006b). Sensitivity and consistency may also exceed those of rabbit assays (Schindler *et al.* 2007). These assays may also provide insights into mechanisms of pyrogenicity and acute pro-inflammatory reactions in patients (Montag *et al.* 2007).

Drug hypersensitivity prediction

Drug hypersensitivity reactions are known to be mediated by covalent binding of drug haptens (partial *antigens*) to proteins. The resultant products are then processed to *peptides* and presented on the surface of MHC (major histocompatibility complex) molecules. These molecules play a key role in immune system activation. The hapten–peptide complexes they present are recognised by T cell receptors (TCRs) that fit the antigen in question (Pichler 2007).

Using blood mononuclear cells, Pichler (2007) showed that direct non-covalent binding of drugs to TCRs is also possible, and may stimulate T cells into dividing and triggering inflammatory responses through cytokine secretion, resulting in cell death (the *p-i concept* – direct pharmacological interaction of drugs with immune receptors).

Such immune-mediated inflammatory responses may cause mild symptoms, such as maculopapular exanthema, through to severe reactions, such as Stevens–Johnson syndrome, toxic epidermal necrolysis, hepatitis, pancreatitis, fever, vasculitis, eosinophilia, and even death (Pichler 2007).

To test the safety of *systemically* (i.e. orally or parenterally) applied drugs that act via this p-i mechanism, human-based *in vitro* assays are necessary, because such immunological reactions are highly species-specific. Small alterations of TCR morphology may dramatically alter T-cell reactivity. Furthermore, reactivity varies even between human individuals, both *in vivo* and *in vitro*, for reasons that are hypothesised to include variations in TCR repertoires, immune regulation, and genetic background. Pichler speculated that a greater understanding of these factors could lead to novel human-based *in vitro* assays for drug hypersensitivity.

Bone-seeking pharmaceuticals

Treatment and monitoring of *osseous* (bony) tumours requires bone-seeking pharmaceuticals, such as radioactively labelled polyphosphonates. Despite their use for more than 30 years, however, uncertainty about the binding interaction between bone-seeking pharmaceuticals and target bone impedes rational drug design of novel bone seekers, such as analysis of SARs. Accordingly, biodistribution and efficacy studies in laboratory animals have traditionally been utilised early in preclinical evaluation (Mitterhauser & Toegel 2008).

Mitterhauser and Toegel (2008), however, described an *in vitro* assay based on simple binding to hydroxyapatite ($Ca_5(PO_4)_3OH$ – a mineral phosphate of calcium that is the principal structural element of vertebrate bone). *In vitro* binding characteristics correlated well with those obtained *in vivo* using mice and *ex vivo* using human donor bone. Accordingly, this assay provides a viable alternative to animal models for the comparative evaluation of novel bone seekers.

Developmental toxicity assessment using embryonic stem cells

Organisation for Economic Co-operation and Development (OECD) Test Guidelines for the assessment of developmental toxicities such as embryotoxicity and teratogenicity are costly and time consuming. Two- or multi-generation studies may consume up to 3,200 animals per substance (OECD Test Guideline 414: *Prenatal Developmental Toxicity Study*), and it is expected that developmental toxicity testing will consume the highest animal numbers in REACH (Pellizzer *et al.* 2005).

The use and further development of appropriate *in vitro* systems is therefore a high priority. These systems may include cell cultures (e.g. embryonic stem cell tests (ESTs)), organ cultures (e.g. micromass assays), and embryo cultures (e.g. whole embryo cultures) (Huggins 2003, Pellizzer *et al.* 2005). Brown (1987) and Brown and colleagues (1995b) provided extensive overviews of existing *in vitro* systems. The ability of ESCs to differentiate into a wide range of potentially vulnerable target tissues maximises their utility. They may be used to test for embryo toxicity *in vitro*, and to screen for teratogenicity and growth retardation (Luttun & Verfaillie 2006). hESCs are preferable, as resultant assays are relatively simple and reproducible, avoid interspecies differences, and may facilitate studies of human development.

Several approaches are possible using ESCs. Traditionally, toxicity has been assessed by morphological analysis of toxic inhibition of beating *cardiomyocytes* (heart muscle cells) in embryoid body outgrowths derived from murine ESCs (Luttun & Verfaillie 2006), although the objectivity of this assay has been criticised (Buesen *et al.* 2004). An immunological assay was described by Seiler and colleagues (2002 & 2004) which used intracellular staining and flow cytometry to detect changes in sarcomeric myosin heavy chain and α-actinin during cardiac differentiation of ESCs. Cultivation of transgenic ESCs with green fluorescent protein, driven by a specific tissue gene promoter, may allow visual detection of teratogenicity (e.g. Paparella *et al.* 2002). Array technologies (see 'Toxicogenomics', pp. 117–18) now allow detection of changes in the expression of thousands of genes simultaneously (Lockhart & Winzeler 2000) following chemical exposure. Analysis of gene expression profiles at different stages of cardiac and skeletal ESC differentiation has been used to identify developmental stages of chemical vulnerability (Pellizzer *et al.* 2004a & 2004b).

An international ECVAM study validated an EST using a set of 20 reference compounds characterised by high-quality *in vivo* embryotoxicity data from laboratory animals and humans. The EST predicted the embryotoxicity of the 20 compounds with 78 per cent accuracy, and 100 per cent predictivity was obtained for strong embryotoxins (Genschow *et al.* 2002 & 2004). Similarly, Whitlow and colleagues (2007) recently conducted ESTs on 6 chemicals with toxicity characteristics established from the literature and this ECVAM study and on 10 Roche internal pharmaceutical substances already tested *in vivo*. Their EST model correctly classified 81 per cent of substances.

Clearly, ESCs offer several avenues for the identification and investigation of teratogenicity in a wide range of differentiated tissues, although

ESTs detect embryotoxicity occurring only early in embryogenesis. However, detection of compounds that show such toxicity is a high priority (Buesen *et al.* 2004). In recognition of this, and of their potential during embryotoxicity testing, investigation of hESCs is included in an important integrated project entitled ReProTect, which is part of the 2002 EU Sixth Research Framework Programme. The ReProTect collaboration includes 35 academic, governmental, and other institutions, and was allocated €9.1 million over five years. Its principal aim is to create new *in vitro* models, and to combine these with existing models in a test strategy aimed at providing information on chemical hazards to the mammalian reproductive cycle (Pellizzer *et al.* 2005).

Considerable standardisation and harmonisation of hESC assays remains necessary, however, along with enlargement of the reference database and consolidation of the details of the existing prediction model, to ensure their regulatory acceptance. The introduction of a metabolic system as an adjunct to the validated EST protocol – which is applicable only for compounds that do not need metabolic activation – will also be necessary (Buesen *et al.* 2004).

Different international approaches to embryo protection and other ethical issues raised by hESC research present significant obstacles (Pellizzer *et al.* 2005). Among other restrictions in the relevant 2003 EC guidelines, for example, is the statement that '[r]esearch will be funded only when there is no adequate alternative available. In particular, it must be demonstrated that one cannot use existing embryonic or adult stem cell lines' (EC 2003b).

Genetic engineering

Incorporation of green fluorescent protein into cells to provide visual confirmation of genetic damage was described above (see 'GreenScreen genotoxicity assays', pp. 110–11). Another particularly important application of genetic engineering is the incorporation of *endogenous* xenobiotic metabolising enzymatic ability into established cell lines. This can enhance the ability to screen for toxicity, and can provide information about molecular mechanisms of metabolism.

Xenobiotic metabolising enzymes have now been expressed in bacterial, yeast, insect, and mammalian cells. More than 20 cell lines, mostly derived from human and hamster tissues, have been genetically engineered for the expression of xenobiotic metabolising enzymes within the past decade. Much of this development has relied on recombinant DNA techniques (Weibel *et al.* 1997). cDNA encoding for nearly every major human P450 isoform has been isolated and sequenced (Crespi 1995).

Recent research has focused on the isolation of the activity of single enzymes within a metabolic profile, followed by study of the activities of multiple enzymes acting in sequence, as occurs *in vivo*. Despite considerable progress in this field, concerns remain about *in vivo* applicability, and much validatory work remains necessary (Huggins 2003).

Toxicogenomics

Toxicogenomics is the application of *genomics* (the study of genomes) to toxicology. Related fields include *transcriptomics, proteomics* and *metabolomics*. Toxicogenomics seeks to link induced gene expression changes with classes of toxic compounds (Farr & Dunn 1999, Aardema & MacGregor 2002). For example, Ellinger-Ziegelbauer and colleagues (2008) identified gene expression profiles characteristic of genotoxic carcinogens (GCs) and non-genotoxic carcinogens (NGCs) using liver samples from rats treated for up to 14 days. On the basis of these results, they were able to predict the carcinogenicity profiles of an independent set of validation compounds with up to 88 per cent accuracy.

cDNA microarrays ('gene chips') contain hundreds or thousands of microscopic spots of cDNA transcripts of messenger RNA (mRNA) templates. They offer the ability to examine many genes simultaneously, and to characterise phenotypic changes in whole cells or organs. This may considerably speed up the generation of results in comparison to less modern methods such as the analysis of transcription (mRNA) by *northern blotting*, the analysis of translation (proteins) by *western blotting*, or the analysis of mutation frequency by *phenotypic selection*. Furthermore, with durations of days rather than years, and with the ability to detect toxic changes much earlier than traditional, more invasive endpoints, cDNA microarrays may facilitate considerable refinement of laboratory animal use (Knight *et al.* 2006c).

Particularly exciting is the ability of cDNA microarrays to detect NGCs (Goodman 2001). After exposing rat hepatocytes to several rodent GCs and NGCs, as well as to two non-carcinogenic hepatotoxicants, during five-day repeat-dose *in vivo* studies, Kramer and colleagues (2004) hybridised fluorescently labelled probes generated from liver mRNA against rat cDNA microarrays. Correlation of the resulting data with the estimated carcinogenic potential of each compound and the dose level identified several candidate molecular markers of rodent non-genotoxic carcinogenicity, such as transforming growth factor-β-stimulated clone-22 and NAD(P)H-cP450 oxidoreductase.

However, microarray technology is still in its infancy, and several existing limitations would benefit from further research and development. Clearly, not all genes can be included in microarrays, so the hybridisation results will represent only a subset of the global changes in gene expression. Care must be taken to ensure that the array chosen is appropriate for the hypothesis under investigation, lest important changes in expression be missed. Despite this limitation, the major challenge of microarray work remains the ability to convert a long list of expression results into an interpretable form, necessitating further research into data analysis (Docterman & Smith 2002).

Nevertheless, it is expected that many of the genes affected by toxicant action will be identified within the next 20 years, leading to progressively greater utility of gene chips as replacements for animal use in toxicity testing (Huggins 2003).

Other human studies

Clinical trials

Clinical studies in human volunteers are an essential component of drug development. They seek to identify differences between the safety and efficacy of pharmaceuticals in preclinical studies and in humans.

In 2006, however, during a UK phase I clinical trial on the anti-inflammatory monoclonal antibody TGN1412, all six volunteers rapidly developed cytokine release syndrome after dosing, culminating in multiple organ failure requiring intensive care. One volunteer suffered permanent damage. These severe adverse reactions occurred despite thorough preclinical trials, which had included computer modelling and tests on human cells, mice, rats, New Zealand White rabbits, marmosets, rhesus macaques, and cynomolgus monkeys, and despite subsequent authorisation by the Medicines and Healthcare products Regulatory Agency and the relevant institutional medical ethics committee. They occurred despite correct drug manufacturing processes and administration in accordance with approved protocols (Bhogal & Combes 2006).

Such cases highlight the lack of certainty in predicting human risks from preclinical studies, and the necessity of compliance with measures designed to minimise those risks. Such measures were discussed in detail by Bhogal and Combes (2006), and at an NIH meeting (Kramer *et al.* 2006).

They include the following:

1. The outcomes of preclinical and clinical studies should be released into the public domain, either once the drug is marketed or when it has been deemed to fail. Such data may guide further research and development, and may influence the design of future clinical trials. Safety and efficacy-based considerations must be correctly acknowledged as outweighing concerns about commercial confidentiality.
2. Greater effort should be invested in the development and use of preclinical test methods more predictive for human outcomes, particularly the use of human cells and tissues in culture, and especially in areas that are currently poorly modelled, such as the immune system (Silliman and Wang 2006). Data from such tests should be considered at least as important as traditional animal test data.
3. The first administration to humans should use microdoses (*phase 0 studies*). The role of microdoses in early drug development was reviewed by the FDA (CDER 2006). *Microdose studies* generally seek to provide information on human pharmacokinetics and bioavailability (Combes *et al.* 2003), rather than pharmacological effects. They use doses estimated from preclinical studies to be well below any thresholds believed necessary to achieve pharmacologic or toxicological activity. However, they may not exhibit threshold or other dose-dependent effects.
4. Increased research should be conducted with the goal of identifying *biomarkers of toxicity*, such as the presence of metabolite activating and detoxifying enzymes (Besaratinia *et al.* 2002), cellular markers of exposure, the presence of mutagens in body fluids, and chromosomal alterations (Cone & Rosenberg 1990). Volunteers should be carefully monitored for the expression of such biomarkers following microdosing, with assessed biological parameters expanded to incorporate possible macroscopic effects as doses are increased.
5. Dosing should be staggered, rather than concurrent, perhaps based on expected *in vivo* half-lives. In the case of TGN1412, adverse clinical signs were evident within two hours of dosing; hence staggering doses by as little as one hour would have significantly reduced the number of volunteers affected.
6. The limited predictivity of human clinical trials for broader populations may result from their focus on small groups of healthy young men or from insufficient study durations. Particularly in phases I and II, small cohorts of young men (20–300) are typically used to

minimise experimental variability, and to eliminate possible endocrinological disruption or other risks to women of reproductive age. Although 1,000–3,000 volunteers may be used in phase III trials – the final phase before marketing (NIH 2006) – it is clear that cohort numbers, study durations, or other aspects of protocol design, conduct, or interpretation are inadequate to detect the adverse side-effects of the large number of pharmaceuticals that harm patients after marketing (see p. 40). Longer studies of more broadly representative human populations would be more predictive.

7.   When drugs are specifically targeted to modulate systemic processes, such as immune system activity, serious consideration should be given to using appropriate patients, rather than healthy volunteers, as soon as physiologically active doses are reached. Immunocompromised patients, for example, might not display the adverse reactions elicited in healthy, immunocompetent volunteers.

Such strategies would probably increase the already substantial time and cost of pharmaceutical development. However, with adverse reactions to pharmaceuticals representing the fourth to sixth leading cause of death in US hospitals (based on a 95 per cent confidence interval; Lazarou & Pomeranz 1998), it is clearly necessary to question the safety for patients of the traditional drug testing paradigm.

Epidemiological, sociological, and psychological research

Despite our extensive tradition of animal testing, little certainty exists about the human toxicity of the vast majority of chemicals in widespread use. Increased *epidemiological* research (the identification of aetiological agents or risk factors through population studies) would enable definitive identification of more human toxins. This would also increase the data set available for validation studies of test models. Concurrent *sociological* or *psychological studies* may increase our understanding of factors predisposing towards risky lifestyle choices.

In some cases, existing social infrastructure can be recruited in support of such studies. For example, cancer treatment centres could be funded to establish tumour registries aimed at identifying new lifestyle, occupational, environmental, and medical carcinogens. Post-marketing surveillance of human pharmaceuticals, with reporting of adverse side-effects, is already required in most jurisdictions and should be fully utilised, perhaps through the provision of incentives to participating physicians, whose available time may be limited.

Additional human investigation

A wide range of other investigative tools are available for the non-invasive examination of human beings. To varying degrees, biomechanical and other bodily systems and processes can be modelled mathematically, and their behaviour predicted by *computerised simulations* (Langley *et al.* 2007). For example, the effects of dental or surgical interventions can be simulated by the incorporation of the biomechanical characteristics of various human tissues into mathematical models. The structural properties of human tooth, bone, and ligament, and of dental materials such as ceramic and titanium, can be incorporated into simulations of dental prosthetic implants to allow predictions of stresses and strains and of the effects on implant movement and fracture probabilities of various mechanical or thermal loads (Gallas *et al.* 2005). Such simulations may use *finite element* (FE) *analysis* – an engineering technique. Structures are modelled by subdividing them into a meshwork of theoretical elements bounded by nodes. The magnitude at various nodes of effects such as strains and stresses applied to the structure can then be computed mathematically. FE analysis has been widely applied in studies of dental materials, oral and maxillofacial surgery, orthodontics, and dental restorations (Mackerle 2004).

The use of *donated human tissue or cell lines* is appropriate for certain biochemical questions, such as those investigating human cellular receptors. In some cases *surrogate tissues* may be used. For example, mast cells isolated from human placenta have some features in common with neurons – including the possession of a functional receptor for nerve growth factor – and may therefore be used for appropriate neuropharmacological, neurophysiological, and neurotoxicological studies (Purcell & Atterwill 1995).

The ability of hESCs to produce a wide range of differentiated cell types makes them particularly valuable. Schrattenholz and Klemm (2007), for example, described models of human neurodegenerative diseases based on neurons derived from hESC precursors. Using calcium imaging (because various neuropathological processes appear to be associated with altered calcium levels), mitochondrial potential measurements, and western blots, they demonstrated their functionality and, further, that the model reproduced crucial mechanistic aspects observed during ischaemia and *excitotoxicity* that are considered central to some neurodegenerative diseases.

Anatomical investigations of donated human cadavers and tissues may be conducted via *gross* or *histological* examination, and a variety of *imaging techniques* may be used in human volunteer studies, such as plain or contrast radiographs, ultrasonography, computed tomography, functional magnetic resonance imaging (fMRI), and positron emission tomography (PET). Such minimally invasive investigative tools may also be applied in animal studies, thereby achieving refinement of animal use.

Even the complexities of the human brain may sometimes be amenable to such study. Neural activity in various regions of the brain can be correlated with specific cognitive, physiological, or physical activities via fMRI or PET. The functional necessity or redundancy of regions thereby identified can be investigated using *transcranial magnetic stimulation*. Applied using a coil, an electromagnetic field disrupts neurological activity in an area of interest for tens of milliseconds or longer, during which volunteers attempt to perform cognitive, visual, or other tasks (Langley *et al.* 2000, Heindl *et al.* 2008).

Integrated toxicity testing strategies

Toxicity testing protocols using tiered combinations of non-animal assays have been described, for example by the UK Department of Health (2000), Worth and Balls (2002), Knight and colleagues (2006c), Combes (2007), and Combes and colleagues (2007). They should include elements appropriate to the individual case, which would normally be conducted in sequence, as in Figure 8.1.

Not all testing stages will be necessary in every case, particularly when positive toxicity data are obtained in earlier stages. The appropriate level of testing will be influenced by factors such as production volumes, human exposure risks, and legislative or regulatory requirements.

Summary

A broad range of investigative tools exist with the potential to replace much animal use in biomedical research and toxicity testing. These include mechanisms to enhance the sharing and *assessment of existing data* prior to conducting further studies, *physicochemical evaluation*, and *computerised modelling*, including the use of SARs and expert systems.

A variety of *tissue cultures* may be used, including immortalised cell lines, embryonic and adult SCs, and organotypic cultures. *In vitro assays* using bacterial, yeast, protozoal, mammalian, or human cell cultures exist for a wide range of toxic and other endpoints. These may be used individually or combined as batteries, and kept static or perfused. *Human hepatocyte cultures* and *metabolic activation systems* may facilitate

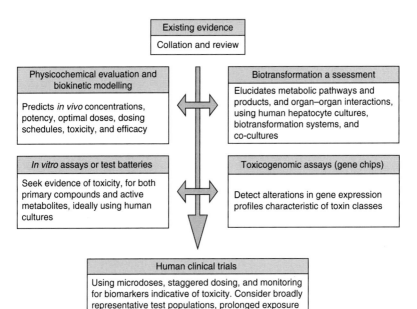

Figure 8.1 Integrated toxicity testing
Source: Knight (2008e).

identification of metabolic pathways, determination of metabolites produced, and assessment of organ–organ interaction. *Microarray technology*
may allow genetic expression profiling of toxins, increasing the speed
with which they are detected, well before more invasive endpoints.

Enhanced *human clinical trials* using microdosing, as well as *surrogate
human tissues, advanced imaging modalities,* and *human epidemiological,
sociological, and psychological studies,* may increase our understanding
of illness aetiology and pathogenesis, and facilitate the development of
safe and effective clinical interventions.

Non-animal investigative methods cannot, of course, provide answers
to all questions about humans, particularly given present technological limitations. However, the same is certainly true of animal models,
which have a more limited capacity for further development.

With respect to toxicity testing, not all appropriate assays have been
scientifically validated or adopted by regulatory authorities. However,
properly collating and examining the more targeted data obtained
through such a testing scheme is likely to yield a weight-of-evidence
characterisation for human toxicity of superior predictivity to that

currently offered by conventional animal-based assays such as the traditional rodent carcinogenicity assay. It may also facilitate greater understanding of human mechanisms of toxicity. Furthermore, non-animal models can offer certain important advantages in comparison to animal use. Particularly when humans or human tissues are used, such alternatives may generate faster, cheaper results that are more reliably predictive for humans, and may yield greater insights into human biochemical processes.

Such considerations are becoming increasingly important given the logistical challenges posed by the high-throughput toxicity testing requirements of US and EU programmes such as HPV and REACH, and also given growing social pressures to find alternatives to laboratory animal use. These pressures are increasingly reflected in legislative or regulatory changes, such as the seventh amendment to the European *Cosmetics Directive 76/768/EEC*, which included detailed provisions on the phasing out of animal testing of cosmetics (Knight 2008e), and the recent *Directive 2010/63/EU on the protection of animals used for scientific purposes*, which strengthens the protection afforded to laboratory animals within Europe (EU 2010).

9
Reduction and Refinement of Laboratory Animal Use

This chapter summarises strategies aimed at achieving *reduction* (decreased numbers) and *refinement* (decreased suffering) of laboratory animal use, and the potential for synergistic or, occasionally, detrimental effects when multiple 3Rs strategies are implemented concurrently.

Reduction alternatives

Strategies designed to achieve reduction of laboratory animal use may be applied at three different levels (De Boo & Hendriksen 2005).

Intra-experimental reduction

Intra-experimental reduction is applied directly at the level of the experiment. Reduction can often be achieved through improvements in experimental design and statistical analysis. Many experiments are poorly designed and use elementary statistics (Fry 2004). Insufficient sample sizes leave many studies underpowered, limiting the statistical validity of study conclusions. Animal lives and resources may also be wasted if experiments subsequently require repetition. The relatively poor statistical knowledge of many animal researchers may be the cause of the high prevalence of poor sample size choices in animal studies. Solutions could include the training of researchers in statistics and the direct input of statisticians into experimental design and data analysis.

Factors that should be considered when calculating appropriate sample sizes include *detectability threshold* (the size of the difference between treatment groups considered significant); known or expected data variation; the required significance of the test (*p-value* or α: the probability of a *type I error* – assuming a difference where none exists); the acceptable probability of assuming no difference where one does exist (β, a *type*

II error; the *power* of an experiment = $1 - \beta$, and 0.8 is the usual choice); and the type of statistical analysis to which the data will be subjected. Smaller detectability thresholds, greater data variation, smaller acceptable error probabilities (greater power), and certain statistical tests for differences all require larger samples.

No universal rule exists for calculating correct sample sizes (De Boo & Hendriksen 2005). Festing (1997), for example, describes two methods: the preferred 'power calculation' and the 'resource equation'. *Power calculations* use formulae which are available in interactive computer programmes (e.g. Van Wilgenburg *et al.* 2003 & 2004). They calculate the minimum sample sizes required to detect treatment effects with specified degrees of certainty. Mead's resource equation (1988) calculates sample sizes using *degrees of freedom*, and incorporates statistical parameters such as treatment effects, block effects, and error degrees of freedom.

Meta-analysis involves the aggregation and statistical analysis of suitable data from multiple experiments. For example, Festing described methods of increasing experimental efficiency through blocking treatments and multifactorial designs (Fry 2004). In *randomised block designs*, animals are randomly allocated to treatment groups in blocks that are designed to be homogeneous other than for the factor being tested. *Nuisance factors* resulting from inevitable differences between blocks are held constant within each block. This decreases data variability, and hence the sample sizes required. Standardisation of conditions within blocks also facilitates comparison with similar blocks that may potentially exist at different laboratories or times, whether or not correction for specific differences is required.

Multifactorial designs introduce multiple factors, such as different sexes or strains, into a single experimental protocol. This allows several experiments to be performed simultaneously, and reveals factor-related effects. For some purposes, treatment and control groups may also be combined, permitting group numbers to be minimised.

By designing experiments and reporting protocols to maximise their utility for later meta-analyses, the benefit of individual controlled randomised experiments can be maximised (Phillips 2005). Although new information can be derived through meta-analysis, more frequently the results allow refinement of existing knowledge.

Pre-screening of test compounds using replacement methods can result in a significant reduction in animal numbers. Analysis of pilot test data – whether or not from animal studies – may yield information about data variance, which can then be used to calculate sample sizes. When information about the size and variability of responses is already available, smaller, more focused studies may become possible. Modifications should also be made in routine tests when opportunities to reduce animal use arise, for example, as a result of an increase in expertise, further standardisation of procedures, or availability of animals with lower inherent variability.

Strategies should also be considered for minimising animal numbers without unacceptably compromising statistical power. Several of these strategies aim to decrease data variability (thereby maximising experimental precision) by minimising heterogeneity in experimental environments and protocols (Eskola *et al.* 1999, Festing & Altman 2002, Balcombe *et al.* 2004, De Boo & Hendriksen 2005, Balcombe 2006). This can be achieved by various means:

1. the appropriate use of environmental enrichment, aimed at decreasing physiological variation resulting from barren laboratory housing and stressful procedures (pp. 30–3);
2. choosing, where possible, to measure variables with relatively low inherent variability;
3. the use of genetically homogeneous (isogenic or in-bred) or specified pathogen-free animal strains; and
4. screening raw data for obvious errors or outliers.

Measurement errors can be reduced via repeated measurements, standardised protocols, improved equipment, and staff training – which may reduce, for example, intra- and inter-observer variation in behavioural assessments (De Boo & Hendriksen 2005).

Strategies such as these, aimed at maximising the statistical power of small samples, are particularly appropriate when marked ethical, cost, or practical constraints limit the number of animals that may be used (e.g. in experiments involving non-human primates).

Finally, the statistical analysis of the resultant data should be closely linked to the experimental design, and to the type of data produced (Festing *et al.* 1998).

Supra-experimental reduction

Supra-experimental reduction involves changing the setting in which experiments occur. Examples include improving the education and training of staff (e.g. training researchers in experimental design and statistics), including a 'named statistician' on animal ethics committees, achieving reductions in breeding surpluses, and critically analysing, and re-defining, test specifications.

Searching for alternatives

Thorough, competent literature searches may reveal a range of alternative strategies, from complete animal replacements to refinement methodologies, that can be applied to proposed experiments. They may even reveal the existence of relevant research under way or completed elsewhere, thereby allowing experimental modification in the light of such new knowledge, and avoidance of unnecessary duplication.

Policy Guideline 12 in the United States Department of Agriculture (USDA) *Animal Care Resource Guide* (USDA 2000b) was developed to support the US *Animal Welfare Act* (*AWA*). It requires principal investigators considering animal research to consider alternatives to procedures that may cause more than momentary or slight pain or distress to the animals, and to provide a written narrative description of the methods and sources used to determine that alternatives were not available, before animal use licences may be granted. All three Rs must be addressed, rather than replacement methods alone.

Accordingly, learning to conduct such literature searches should be a key part of the training of all laboratory animal scientists. The USDA Animal Welfare Information Center (http://awic.nal.usda.gov/) and several other sources provide helpful guidance on literature searching and alternatives. De Boo and Hendriksen (2005) suggested that doctoral students using animals should be required to include justifications for animal use in their dissertations, including an explanation of why alternative methods were not used.

Retrospective analysis

Experimental protocols frequently stipulate animal numbers with a high degree of arbitrariness, particularly where regulatory testing is required. Hence, as stated, numbers may sometimes be in excess of, or less than, those required to achieve statistical significance. In addition, protocol modification over time through standardisation or improvement of experimental procedures or expertise may decrease experimental variability, and hence animal numbers required. Hence, regular *retrospective analysis* of test data is advisable to indicate changes in data variance and experimental numbers required. Through such retrospective analyses, Hendriksen and colleagues (1987) and Knight and Roberts

(1987) showed that, for tetanus potency testing routinely performed in their institutes, a reduction of about 25 per cent in the number of animals used would be possible, with only a minor effect on confidence interval ranges.

Lindl and colleagues (2005) demonstrated a profound discordance between actual experimental utility and claims of expected human benefit made on licensing applications (see 'Experiments expected to yield medical advances', Chapter 5, p. 50). Retrospective evaluation of animal experiments could assess the degree to which experimental objectives were successfully met, and also the extent to which animal suffering matched that predicted in licensing applications. Accordingly, such retrospective evaluation should be routine, to inform both future research design and future experimental licensing decisions.

Animal re-use

Policies encouraging the *re-use* or *recycling* of animals (the so-called fourth R) represent a special case of supra-experimental reduction. In some *longitudinal studies*, for example, animals may act as their own controls. This can make the use of parallel control groups, as well as interim kills, unnecessary. In some studies involving animal re-use, however, the impact on such animals may be increased, and so a weighted case-by-case cost/benefit assessment must be made to ensure that such impacts do not exceed the benefits achieved by minimising numbers. It may, for example, result in higher animal welfare and ethical standards when a lower impact on a larger number of animals is incurred, rather than a severe impact on a few (De Boo *et al.* 2005). Both the AWA and European *Directive 2010/63/EU on the protection of animals used for scientific purposes* prohibit re-use where this results in the substantial accumulation of pain and suffering (US Government Printing Office 1985, EU 2010).

At the departmental level, or even across departments within institutions, good planning of research activities, cooperation, and sharing of animals and resources between research programmes may further decrease the numbers of animal used.

Extra-experimental reduction

Extra-experimental reductions are achieved through more distantly related developments, such as improved research or production strategies, that result in superior quality, consistency, and safety of biological products or pharmaceuticals, thereby decreasing data variability and test group sizes, as well as the perceived need for animal testing.

Discretionary toxicity testing

Traditionally, toxicological assessment has consisted of four consecutive phases: hazard characterisation, hazard assessment, exposure assessment, and, finally, risk assessment. Consequently, the toxicological profile of each chemical has been evaluated even if it becomes clear on completion of the process that the likelihood of exposure is so low that toxicity is not a realistic risk for workers or consumers. Accordingly, the *reverse toxology approach* proposes first assessing exposure levels to determine the extent of further assessment and testing necessary. Similar recognition of decreased testing requirements has allowed the replacement of the classical *LD50 test* with OECD-recognised tests such as the fixed dose procedure, the acute toxic class method, and the up-and-down procedure, which are considered to yield sufficient indication about toxicological hazard when exposure risks are lower. By reducing animal numbers and impacts on test subjects, such modernised procedures facilitate both reduction and refinement of laboratory animal use.

Similarly, the REACH chemical testing system uses a tiered testing system: chemicals produced or imported in the greatest quantities are subjected to the most comprehensive testing, while chemicals in lower tiers have reduced testing requirements, including for animal testing.

Modernisation of polyclonal antibody production

In some cases, such as the production of biological products, modernisation of laboratory protocols may result in reduction of animal use. For example, polyclonal antibodies (PABs) are often used in diagnostic testing and immunological studies. Traditionally, these products are produced by immunising rabbits or mice with the antigen of interest, and bleeding the animals several weeks after the last immunisation. In some cases, however, particularly when antibodies are required to highly preserved mammalian antigens, it might be preferable to obtain PABs from the egg yolk of an immunised chicken. This facilitates reduction of animal use, because 1 chicken supplies as many PABs as approximately 10 rabbits, as well as refinement of laboratory animal use, because blood sampling becomes unnecessary (Schade *et al.* 2005).

Standardisation of good laboratory practice

The development of quality-controlled manufacturing processes for the production of biological products such as vaccines provides a good example of extra-experimental reduction. In this case GMP and QA principles, which include testing at various checkpoints in the production chain, aim to ensure the consistency of each batch of products, thereby decreasing the safety and efficacy testing required for the final products (Hendriksen 2006, Knight 2008e).

GMP and QA principles may be used by pharmaceutical companies to ensure the quality, safety, and efficacy of the medicines or biological products produced. Related good laboratory practice (GLP) consists of administrative procedures, including standard operating procedures, governing the organisation of test facilities and the conditions under which studies are conducted. A key aim of such procedures is to ensure the generation of reliable high-quality data suitable for assessment by regulatory agencies, thereby minimising data variability and the need for repeat testing.

Most countries have issued regulations relating to GLP and GMP, and aim to ensure compliance through laboratory inspections and data audits. The implementation of such standards also facilitates the international acceptance of data and the harmonisation of test protocols.

Harmonisation of test guidelines

Regulations for product licensing are issued by national and international regulatory bodies such as the EC, FDA, and EPA. Unfortunately, such regulations and underlying test programmes and specifications frequently differ between regulatory bodies. Consequently, companies that wish to market their products in several countries must comply with multiple test requirements, which frequently results in duplicative or additional animal testing. The numbers of animals used can be large, and such testing increases the costs of product development.

Accordingly, several activities have been initiated to harmonise such regulations. The International Conference on Harmonisation of Technical Requirements for Registration of Pharmaceuticals for Human Use is a cooperative endeavour between the regulatory authorities of Europe, Japan, and the US, and the pharmaceutical industry. It is aimed at achieving greater harmonisation in the interpretation and application of requirements for product registration, and at reducing or obviating duplicative animal testing. This is also a strategy by which obsolete animal studies can be replaced by more advanced techniques, including appropriate non-animal methodologies.

The Pharmacopoeia Discussion Group represents the same interest groups, and has similar aims. Unfortunately, the rate of international harmonisation of test guidelines is very slow. An alternative approach is to seek more limited agreements on mutual recognition and acceptance of test data.

Refinement alternatives

Experimental refinements include the use of analgesic and anaesthetic modalities that will not unduly alter experimental outcomes of importance. Such modalities remain underused, although it appears likely that a substantial minority of procedures internationally are markedly invasive (Chapter 3, pp. 22–6). Over the past two decades, British procedures not utilising any form of anaesthetic fluctuated between 59 and 69 per cent, and reached 67 per cent in 2009 – the most recent year for which data were available at the time of writing. Analgesic use was not recorded (Chapter 3, p. 26).

While the use of anaesthetics or analgesics undoubtedly alters normal physiology, claims that such alterations are sufficiently important to hypotheses under investigation to warrant their exclusion require careful scrutiny. It may, for example, be the case that physiological variations in response to painful stimuli are of greater consequence.

It is also important that researchers and technicians handling animals minimise the stress of their subjects. The Federation of European Laboratory Animal Science Associations has identified four categories of professionals who should receive education and training (Nevalainen *et al.* 1999):

1. those taking care of animals;
2. those carrying out animal experiments;
3. those responsible for directing animal experiments; and
4. laboratory animal science specialists.

Benefits include enhanced animal welfare, and decreased data variability and scientific error resulting from animal stress or operator inconsistency. Accordingly, training in laboratory animal handling and husbandry is of paramount importance, and such education and training is now mandatory in some EU countries. In the Netherlands, for example, only scientists with an academic degree in one of the biomedical disciplines who have completed a three-week course in laboratory animal science are permitted to conduct animal experiments.

Other refinement strategies include the use of non-invasive imaging modalities; the use of *telemetric devices* to obtain information remotely – although such devices may themselves compromise welfare due to surgical trauma and the weight of the device inside or on the animal's body, so case-by-case assessments are necessary; faecal analysis (e.g. faecal cortisol can be analysed, in some cases eliminating the need for *venipuncture*); positive reinforcement techniques such as training animals (especially primates) to participate in monitoring (e.g. presenting arms for venipuncture), rather than relying on physical or chemical restraint; environmental enrichment; and the opportunity to socialise whenever possible, with compatible conspecifics, for social species (De Boo *et al.* 2005).

Interaction between alternatives

In the development of research questions and designs it is advisable to consider all alternative methods for specific procedures together, rather than in isolation. There are instances in which the use of one of the three Rs in tandem with one or both of the other two Rs can work to produce synergistic effects (De Boo *et al.* 2005). For example, introducing education and training programmes for staff is likely to improve animal care, assist with the early detection of welfare problems, and aid in the monitoring of experimental effects. It may result in decreased animal stress as well as improved recording of results, which may, in turn, reduce result variability, and hence animal numbers required.

Similarly, the replacement of toxicity tests on live animals by animal tissue or cell cultures not only reduces animal numbers, but can also minimise the harm experienced by those animals that are used. Reduced costs and timeframes, and increased human predictivity, may sometimes result – particularly if human cell or tissue cultures are used. In such cases all three Rs may converge, to the benefit of science, industry, consumers, and animals.

When implementation of one of the three Rs would conflict with another potentially beneficial alternative method, researchers and ethical review committees need to evaluate the relative impacts of each method. They may be required to make difficult decisions concerning the prioritisation of alternative methods. For example, as a trade-off for replacing live animal use, foetal bovine serum is commonly used in *in vitro* assays, but the collection of blood may cause suffering to the foetus. In this case the collection method could be modified, or a serum-free culture could be used, which offers additional scientific benefits (Even *et al.* 2006). Another example of a conflict between reduction (if increased data acquisition allows the use of fewer animals) and refinement may arise from the use of telemetric devices, which, as mentioned previously, can occasionally produce welfare problems.

Summary

Strategies designed to achieve the reduction of laboratory animal use may be applied at three different levels. *Intra-experimental reduction* is applied directly at the level of individual experiments, and may often be achieved through improvements in experimental design and statistical analysis.

Supra-experimental reduction involves the development and implementation of best practice policies at institutions where animal experiments are conducted. Examples include improving the education and training of staff, including a 'named statistician' on animal ethics committees, critically analysing test specifications, and achieving reductions in breeding surpluses. *Retrospective evaluation* of animal experiments is important to examine the concordance of actual animal suffering with that predicted in licensing applications, and the degree to which experimental objectives were successfully met. This may inform future experimental licensing decisions and improve protocol design.

Animal *re-use* or *recycling* (the 'fourth R') represents a special case of supra-experimental reduction. However, the impact on such animals may be increased, so a case-by-case cost/benefit analysis is necessary to ensure adverse impacts on individuals do not exceed the benefits achieved by minimising numbers.

Extra-experimental reductions are achieved through more distantly related developments, such as the international harmonisation of testing requirements and improved production of biological products such as vaccines – thereby decreasing data variability and the sizes of test groups needed.

The *reverse toxicology approach* implemented in the REACH chemical testing programme provides another example. Likely human exposure levels are assessed first, rather than hazard (the more traditional approach), to determine the extent of further assessment and testing necessary. This constitutes an example of *discretionary toxicity testing*.

Experimental *refinements* include increased use of analgesic and anaesthetic modalities, which remain underused. Others include the use of non-invasive imaging modalities, telemetric devices, positive reinforcement techniques, environmental enrichment, and socialisation opportunities.

Each of the three Rs should be considered in the design of research protocols. In many cases simultaneous implementation of two or more 3Rs methods can synergistically improve both animal welfare and scientific quality. However, conflicts in a small minority of cases may necessitate careful evaluation of the relative impacts of different methods (De Boo & Knight 2008).

Part IV

Educational Animal Use and Student Impacts

10

Educational Animal Use

Animal use in veterinary education

Few, if any, educational disciplines have stronger justifications for invasive animal use than veterinary education. Veterinarians must be familiar with the clinical signs of diseases, which might be proposed as a justification for inflicting those diseases on laboratory animals. They must be able to perform a variety of clinical and surgical procedures, and be able to function professionally when presented with grievously injured animal patients, during surgery, and when euthanasing terminally ill or injured patients. These needs might respectively be proposed as reasons for injuring animals and presenting them to veterinary students, or for requiring students to practise surgical procedures or to kill animals. Hence, in any critical assessment of the necessity of invasive animal use in biomedical education, it is informative to examine closely the discipline in which such use is most strongly justified: veterinary education.

Animal use resulting in harm or death has traditionally played an integral role in veterinary education. Many thousands of animals have been killed worldwide during attempts to teach practical skills or to demonstrate scientific principles which have, in many cases, been established for decades. Animals are killed and dissected to demonstrate anatomical principles. Living animals or organs taken from them are subjected to invasive experiments in physiology, biochemistry, pharmacology, and parasitology laboratories. Veterinary students in most countries learn surgery by practising surgical procedures on healthy animals. Animals surviving these experiments or procedures are usually killed afterwards by the students.

The assertion that animals are harmed in veterinary education is controversial. However, *markedly invasive* procedures, such as surgical and some experimental procedures, do normally incur harm when conducted on healthy animals that do not benefit from them. Harm accrues from any pain, discomfort, or psychological distress associated with the procedure, and from impediments to physical function and disruption of the animal's normal life – all of which are likely to interfere with the achievement of goals important to the animal.

Animal shelter sterilisation programmes, in which homeless animals are neutered by students under supervision, constitute a special case in veterinary education. While individual animals rarely experience immediate benefits from being neutered (although long-term health benefits may accrue), the potential for future suffering is often decreased through prevention of over-breeding and subsequent competition for homes and survival resources, which benefits the population overall.

Some consider that the killing of a healthy animal, when conducted without the infliction of distress or other suffering, does not necessarily constitute harm (e.g. Luy 1998). To varying degrees, deaths approaching such an idealised state are reasonably common in veterinary education, for example, when animals are killed with minimal pain or distress prior to laboratory exercises, or following experiments or surgical procedures conducted under general anaesthesia. However, animals have a broad range of natural interests they seek to fulfil during their lives. The interest in achieving a state of positive physical and psychological well-being is but one example. Death prevents the achievement of almost all the interests of any animal, and therefore constitutes one of the most profound harms that can, in fact, be inflicted (Balluch 2006).

A very rare exception occurs when there is an overwhelming interest in avoiding severe and intractable suffering caused by illness or injury. In this case physical and psychological well-being is elevated from a profoundly negative to a 'null' state through death. This improvement constitutes genuine *euthanasia*, that is, a 'good death' – one that is clearly in the animal's best interests.

Humane teaching methods

During the past two decades there has been a large increase in the development and availability of non-harmful teaching methods such as computer simulations, high-quality videos, ethically sourced cadavers, preserved specimens, models and surgical simulators, non-invasive *demonstration experiments* conducted on students, and supervised clinical experiences (e.g. Bauer 1993, Knight 1999, Gruber & Dewhurst 2004). Some of these warrant further explanation.

Ethically sourced cadavers

Ethically sourced cadavers are those obtained from animals that have been euthanased for medical reasons, or, less commonly, that have died naturally or in accidents. At least nine US veterinary schools and some international veterinary schools have established client donation programmes in their teaching hospitals to facilitate the use for teaching purposes of cadavers from animals euthanased for medical reasons (Knight 2007c).

The first cadaver donation programme was implemented at the Tufts University School of Veterinary Medicine (now the Tufts University Cummings SVM) in 1996 (Kumar *et al.* 2001), and is perhaps the most successful. Clients and clinicians in the Tufts programme make decisions to euthanase on normal medical grounds, but once the decision is made, the euthanasia fee is waived. Because this occurs after the euthanasia decision, financial influences are minimised. Humane euthanasia information leaflets help clinicians explain body disposal options to clients; however, cadavers are not allocated to specific teaching areas until afterwards, and precise details of cadaver use – which may be sensitive – are generally not provided.

Kumar and colleagues (2001) reported that some reluctance of anatomy faculty members to initiate similar programmes elsewhere appeared to be based on assumptions that the operation and administration of such programmes would be labour intensive, and that the high prevalence of neutered animals donated would impede the teaching of reproductive anatomy. However, the client donation programme at Tufts is no more labour intensive than the procurement and embalming of animals from traditional sources. Furthermore, Kumar and colleagues described the ease of recruiting and training students to perform the embalming, thereby saving faculty time. They also reported that a few sexually intact animals are donated each year, which are preserved and re-used, allowing sufficient study of reproductive anatomy at Tufts.

Several advantages accrue from the use of ethically sourced cadavers. These include financial savings; greater biological diversity among specimens; integration of clinical histories, pathological conditions, and ethical considerations into the first year of the veterinary curriculum; and elimination of student and faculty objections to the use of purpose-killed animals (Kumar *et al.* 2001, Fearon 2005).

In 2000, the Tufts University SVM had a caseload of 21,484 dogs and cats. Approximately 240 animals per month (with a canine to feline ratio of about 2:1) were euthanased at the request of the clients, of which approximately 20 were donated to teaching programmes (around 8 per cent). This was sufficient to meet all cadaver needs for educational purposes, including the first-year gross anatomy course, and the clinical skills and medical procedures laboratories. Kumar and colleagues consequently concluded that a similar donor programme should be logistically feasible at any normal veterinary teaching hospital (Kumar *et al.* 2001). By 2005 the Tufts annual caseload had risen to around 31,000 small animals (Fearon 2005).

Humane surgical training

Humane veterinary surgical courses ideally comprise several stages. Students may commence by learning basic manual skills, such as suturing and instrument handling, using knot-tying boards, plastic organs, and similar models. They may then progress to simulated surgery on ethically sourced cadavers. Finally students observe, assist with, and then perform under close supervision necessary surgery on real patients that actually benefit from the surgery – as distinct from healthy animals that are later killed – similar to the manner in which physicians are trained (Knight 1999, Hart *et al.* 2005).

On the face of it, traditional *terminal* surgical laboratories appear to offer the advantages of guaranteed access to, and consistency of, practical training. However, well-designed alternative surgical programmes consistently appear to offer the necessary depth and breadth of surgical experience through a combination of internal and external rotations in veterinary school teaching hospitals, private clinics, and animal shelters. Pavletic and colleagues (1994), for example, described an alternative small animal medical and surgical procedures course developed at the Tufts University SVM. The use of ethically sourced cadavers during the third-year laboratory programme was supplemented with additional clinical training during the fourth and final year, comprising four supplemental weeks in the small animal surgery rotation, and one week in each of the small animal medicine and intensive care rotations. The positive educational outcomes of this and several other alternative surgical programmes are reviewed later in this chapter on pages 156 to 158.

Hart and colleagues (2005) described the trend towards greater clinical and surgical exposure in alternative veterinary surgical curricula. The Faculty of Veterinary Medicine, University College Dublin, for example, requires students to complete at least 24 weeks of extramural practice (external rotations) during the third, fourth, and fifth years of the course, provided at various branches of the practising profession, including state veterinary research and diagnostic institutes (Doherty & Boyd 2006). The University of Cambridge Veterinary School, the

University of Glasgow Faculty of Veterinary Medicine, the University of London Royal Veterinary College, and the University of Sydney Faculty of Veterinary Science are a few of a growing number of veterinary schools that have implemented lecture-free final years to allow students full-time experience in clinical environments such as university teaching hospitals and selected private veterinary practices (Dale *et al.* 2003, Jefferies 2003, May 2003, Canfield & Taylor 2005). Clinical exposure may also begin earlier. At the University of California (Davis), for example, clinical exposure now commences in the first year of the curriculum (Hart *et al.* 2005).

The increased clinical exposure common in alternative surgical programmes provides more realistic educational experiences (Gruber & Dewhurst 2004), offering a myriad of learning opportunities not found outside clinical settings (Smith & Walsh 2003). In particular, external rotations are more likely to expose students to a higher volume of commonly encountered conditions than teaching hospitals, with their invariably higher proportion of referral cases (Kopcha *et al.* 2005). Resultant benefits include greater exposure to the clinical histories, examinations, and presenting signs of cases more directly relevant to new graduates, and to the diagnostic workups and post-operative management of such cases. Surgical participation is usually conducted under close individual supervision, as distinct from the group supervision normally provided during veterinary school surgical laboratories. On the other hand, surgical supervision in veterinary schools is provided by specialists.

Participating external practices are chosen on the basis of factors such as geographical proximity, sufficient caseload, interest in teaching senior veterinary students, willingness to liaise with veterinary faculty, and compatibility of the species caseload, practitioner personalities, and practice facilities with the needs and interests of particular students. Placements are arranged by faculty coordinators or by students. Assessment of experiences typically occurs via a combination of practitioner performance evaluations and student written assignments, such as clinical case logs, case reports, and practice management assessments. Published experiences demonstrate that sufficient participating practices can be successfully identified under normal circumstances to service the needs of major veterinary schools (e.g. Kopcha *et al.* 2005).

Animal shelter sterilisation programmes, in which homeless animals are neutered by students under supervision and returned to shelters, constitute an important part of humane veterinary surgical courses worldwide. The popularity of these programmes stems in part from the fact that all parties benefit from them. The neutered animals are more likely to be adopted (Clevenger & Kass 2003), the number of unwanted animals killed due to uncontrolled breeding and a subsequent lack of homes is decreased, the students gain invaluable experience of some of the most common procedures they will later perform in practice (Richardson *et al.* 1994, Howe & Slater 1997), and the veterinary school experiences the public relations benefits of providing a valued community service (Knight 1999).

Faculty opposition to humane teaching methods

Despite their potential benefits, since at least 1986, it has been my experience and that of veterinary student and faculty colleagues around the world that many veterinary faculty members remain opposed to the introduction of more humane teaching methods.

Australian veterinary opposition

As a veterinary student in 1998 at Western Australia's Murdoch University, I was forced to initiate legal action and mass media exposure of curricular animal killing before Murdoch would allow me to use humane teaching methods in certain units. To its credit, Murdoch then responded affirmatively, by introducing Australia's first formal university-wide policy allowing conscientious objection by students. Murdoch agreed to provide such conscientious objectors with alternatives to harmful animal use during teaching or assessment activities (Knight 1999). Similar policies have since been adopted by at least two other Australian universities (University of Sydney Faculty of Veterinary Science, University of Wollongong), and by several US universities (e.g. University of California (Berkeley), Cornell University, University of Illinois, and Virginia Commonwealth University).

Veterinary students at Australia's other long-established veterinary schools have experienced similar difficulties when seeking to use humane teaching methods. The University of Sydney Faculty of Veterinary Science adopted very progressive policies with respect to humane alternatives in 2000 (elimination of all terminal surgical laboratories, implementation of a pound dog sterilisation programme, adoption of a policy allowing student conscientious objection), but even there students have still faced difficulties when requesting humane teaching methods more recently (Anon. 2006).

Since 1999 I have also corresponded with students seeking to overcome considerable faculty opposition to requests for humane teaching methods at the University of Melbourne Faculty of Veterinary Science and the University of Queensland School of Veterinary Science. All these students were ultimately successful, with the result that by 2005 the first students had graduated from all four established Australian veterinary schools without killing animals during their surgical training.

International veterinary opposition

Reports from veterinary students in the US and elsewhere (described in Knight 2007c) indicate that although a growing number of veterinary schools worldwide have implemented humane teaching methods, such opposition to their implementation remains common. Faculty opposition to student requests for humane teaching methods is an international, rather than an isolated, problem, and is evident in some of the world's leading veterinary schools.

On rare occasions this opposition is described in veterinary journals. Fearon (2005), for example, describes an interview with Professor Kumar, head of veterinary gross anatomy at the Tufts University Cummings SVM. Kumar established Tufts' pet body donation programme in 1996 to facilitate the ethical sourcing of the cadavers of animals euthanased for medical reasons, to replace the use of purpose-killed animals in anatomy dissection and clinical skills training (Kumar *et al.* 2001). Kumar described the opposition of almost all his academic peers at other veterinary schools to student requests for the establishment of similar programmes as 'arrogant', and stated that the general attitude to requests of this sort is that 'you don't let the inmates run the asylum' (Fearon 2005).

In 2002 a *Petition for Rulemaking and Enforcement under the Animal Welfare Act to Eliminate Violations of the Review of Alternatives Provision* was filed by the US Association of Veterinarians for Animal Rights (now the Humane Society Veterinary Medical Association) and several veterinary students with the USDA, which is responsible for administering and enforcing this federal Act. All US veterinary schools were subsequently inspected, and nearly every school was cited for non-compliance with the Act. Most citations were issued for failing to search for alternatives to harmful or lethal animal use, or for failing to provide an adequate explanation as to why non-harmful alternatives were not being used. Many schools were also cited for duplicative use of animals and for the number of animals used, as well as for inappropriate species choice. Some were cited for lack of personnel training and animal identification, for conducting multiple potentially painful procedures, and for lack of information regarding anaesthetics and methods used to kill animals (Anon. 2004b).

International non-veterinary opposition

Such opposition to the use of humane teaching methods is not unique to veterinary educators. Non-veterinary students requesting humane teaching methods have similarly faced strong faculty opposition in a range of institutions internationally (described in Knight 2007c). Such opposition was demonstrated by the prestigious US National Association of Biology Teachers, which at first endorsed the use of humane alternatives in education, but which later rescinded this policy because of opposition from biology teachers. Commenting on this reversal, Van der Valk and colleagues (1999) stated: 'Often, they are not interested in the ethics of using animals. Textbooks, laboratories and equipment are still oriented towards animal experimentation. Convincing these teachers of the advantages and ethics of using alternatives is difficult, the situation being very much polarised. Incorporating the principles of the 3Rs into teachers' initial training and post-qualification professional development would help to overcome some of these difficulties.'

Causes of faculty opposition

Psychological factors

There may be some interesting psychological phenomena underlying the resistance demonstrated by certain faculty members to the use of humane teaching methods, including a personal need to justify the large-scale killing of animals for courses within their responsibility. Gruber and Dewhurst (2004) further asserted: 'Human vanity...should not be underestimated. For many university teachers it is not acceptable to diverge from the methods one was taught and which one has always used in a life of teaching. Aversion to accepting alternatives that were not developed in one's own country also plays a role.'

Appeals to academic freedom to support opposition to humane teaching methods were rebutted in detail by Rutgers University (New Jersey) law professors Francione and Charlton (1992), who described the legal basis for student conscientious objection. Although important, academic freedom is not unlimited, they asserted. Instructors may require students to consider and discuss academically relevant material with which they are uncomfortable, but they may not normally require students actually to participate in acts to which they have sincere and conscientiously held objections. Although Francione and Charlton referred to US constitutional and legislative provisions, the general legal principles described are also applicable to many other countries.

Concerns about resource constraints

The freedom to exercise conscientiously held beliefs is considered sufficiently important under the US Constitution that exceptions are permissible only if it can be proven that accommodating such beliefs – for example, through the provision of humane teaching methods – would impose financial or administrative burdens so severe as to seriously threaten the continued operation of the institution (Francione & Charlton 1992). Unsurprisingly, such arguments have not been successfully made at any educational institution. Concerns have been raised about the time and cost burdens humane teaching methods might incur; however, existing studies demonstrate time and cost benefits, rather than disadvantages, associated with these methods (e.g. Rudas 1993, Dewhurst & Jenkinson 1995, Dhein & Memon 2003). Indeed, these advantages, particularly those of a financial nature, are the strongest drivers of the introduction of humane teaching methods into courses internationally.

Concerns about teaching efficacy

The only argument that might successfully allow the denial of humane teaching methods in a legal or similarly rational forum would be that harmful animal use was truly essential for the acquisition of the skills or knowledge required for the practice of the profession in question. And indeed, in my experience and the experience of veterinary and non-veterinary student colleagues worldwide, the reasons most commonly cited by faculty members opposed to the introduction of humane teaching methods are concerns about their educational efficacy. Given the prevalence of such concerns, reviews of relevant educational studies are warranted, and provide the focus for the remainder of this chapter.

Studies of educational efficacy

Patronek and Rauch (2007) systematically reviewed learning outcomes achieved via humane teaching methods in comparison to those achieved by terminal live animal use. Seventeen studies were retrieved, of which five examined veterinary students, three examined medical students, six examined other university students, and three examined high school biology students.

In two of these studies of medical students, equivalent learning outcomes were achieved using alternatives to the dissection of human cadavers, and harmful animal use may not have occurred (Jones *et al.* 1978, Guy & Frisby 1992). Of the remaining 15 studies clearly involving comparisons with harmful animal use, 4 showed superior and 11 showed equivalent learning outcomes when humane teaching methods were used. Of the five veterinary student studies, two showed superior surgical skills acquisition when alternatives to terminal live animal use were employed, and three showed equivalent learning outcomes when alternatives to harmful animal use were employed in surgical and physiology courses.

Consequently, Patronek and Rauch concluded that 'alternatives are a viable method of instruction in the field of biomedical education'. They encouraged 'biomedical educators to consider how adopting alternative teaching methods could be of benefit to their teaching programs, students, and faculty members'.

By publishing one of the first such systematic reviews, Patronek and Rauch made a major contribution to this field. However, they examined only terminal live animal use associated, for example, with animal dissection; live animal surgery; and live animal physiology demonstrations. Other potentially harmful procedures were excluded from consideration, such as equine or bovine nasogastric intubation conducted by novice practitioners, repetitive equine or bovine rectal palpation, or even potentially stressful confinement and observation of non-domesticated species. Furthermore, only one bibliographic biomedical database (PubMed) was searched for papers published from 1966 to 2004, and the search terms used were somewhat limited. Additional relevant comparative studies of student learning outcomes exist. Consequently, in 2006 I conducted a more comprehensive systematic review of student learning outcomes achieved via humane teaching methods in comparison to those achieved by harmful animal use (Knight 2007c).

For the purposes of this review, animal use considered harmful included:

- invasive procedures, or those reasonably likely to be significantly stressful, such as:
  - equine nasogastric intubation (when conducted by novice practitioners);
  - most physiology, pharmacology, and biochemistry demonstration laboratories using live animal subjects or living tissue from recently killed animals; and

– surgical procedures other than those described below;
- any use of animals resulting in death, other than genuine euthanasia performed solely for medical or severe and intractable behavioural reasons;
- the dissection of purpose-killed animals.

Animal use considered non-harmful included:

- observation of wild or feral animals in field studies, or companion animals;
- minimally invasive or stressful procedures conducted on living animals, such as bovine rectal palpation (although repeated use in some veterinary practical classes can become stressful and/or harmful);
- invasive procedures conducted for the benefit of genuine animal patients or populations, such as neutering operations and similarly beneficial elective surgeries performed on healthy animals, and emergency surgeries conducted on injured or unwell animals;
- dissection, clinical, or surgical procedures performed on cadavers obtained from animals that had been euthanased for medical reasons, or that had died naturally or in accidents (ethically sourced cadavers, similar to the use of human cadavers donated for medical education).

With respect to studies of veterinary surgical training in which surgery performed on living animals was compared with that conducted on cadavers or inanimate models, the source of the cadavers was unspecified in most studies. However, cadavers are usually obtained from ethically questionable sources, such as the greyhound racing industry and animal control agencies (council pounds). Consequently, when cadaver use was compared with a non-animal alternative (e.g. Griffon *et al.* 2000), the non-animal alternative was considered the more humane option for the purposes of this review.

However, cadavers may also be ethically sourced, and a growing minority of veterinary schools have established client donation programmes in their teaching hospitals, as described under 'Ethically sourced cadavers' earlier in this chapter on pages 141 to 142. Given the potential for ethical sourcing, in comparisons with terminal live animal use (the norm in veterinary surgical training) a cadaver was considered the more humane option.

Efficacy of humane teaching methods

Increasing numbers of veterinary schools internationally have introduced non-harmful teaching methods, which have sometimes been accompanied by educational evaluations. Twelve papers published from 1989 to 2006 described studies of veterinary students that compared learning outcomes generated by humane alternatives with those achieved by traditional harmful animal use (Table 10.1).

Greenfield and colleagues described the same study in two references (1994 & 1995); hence 11 distinct studies of veterinary student learning outcomes were retrieved. Nine of these veterinary student studies assessed surgical training – historically the area of greatest harmful animal use.

In 45.5 per cent of cases (5/11), superior learning outcomes (superior skill or knowledge, or equivalent performance with reduced activity times) resulted from the use of the humane option; equivalent learning outcomes also resulted in 5 cases, and in 1 case (9.1 per cent) the humane option resulted in inferior learning outcomes.

Twenty-one papers published from 1968 to 2004 described studies of non-veterinary students in related academic disciplines, similarly comparing learning outcomes generated by humane alternatives with those achieved by traditional harmful animal use (Table 10.2).

Seven of these studies of related non-veterinary disciplines examined high school biology students, and 14 examined university biology, medical, nursing, pharmacology, physiology, and psychology students. The seven studies of high school biology students published from 1968 to 2004 examined anatomical knowledge gained using alternatives to the dissection of purpose-killed animals. Three studies demonstrated superior, three studies demonstrated equivalent, and one study demonstrated inferior knowledge acquisition when humane alternatives were used.

Of the 14 studies examining university students, which were published between 1983 and 2001, 35.7 per cent (5/14) demonstrated that students using humane alternatives achieved superior learning outcomes, or achieved equivalent results more quickly, allowing time for additional learning; 57.1 per cent (8/14) demonstrated equivalent educational efficacy, and only one study (7.1 per cent) demonstrated inferior educational efficacy of humane alternatives.

Twenty-nine papers published from 1983 to 2006 not involving comparisons with harmful animal use were also identified which illustrated additional benefits of humane teaching methods in veterinary education (Table 10.3).

Table 10.1 Veterinary student outcomes: humane teaching methods compared with harmful animal use

Study	Veterinary discipline	Humane option	Total students (humane option)	Humane method superior	Equivalent learning outcomes	Humane method inferior
1 Abutarbush et al. 2006	Clinical skills (equine)	CD-ROM	52 (27)	✓		
2 Bauer et al. 1992	Surgery	Cadavers	24		✓	
3 Carpenter et al. 1991	Surgery	Cadavers			✓	
4 Fawver et al. 1990	Physiology	Interactive videodisc	85	✓		
5 Greenfield et al. 1994	Surgery	Soft tissue organ models	36		✓	
6 Greenfield et al. 1995	Surgery	Soft tissue organ models	36		✓	
7 Griffon et al. 2000	Surgery	Plastic models	40 (20)	✓		
8 Johnson & Farmer 1989	Surgery	Models		✓		
9 Olsen et al. 1996	Surgery	Fluid haemostasis models	40 (20)	✓		
10 Pavletic et al. 1994	Surgery	Cadavers	48 (12)		✓	
11 Smeak et al. 1994	Surgery	Hollow organ simulators	40 (20)			✓
12 White et al. 1992	Surgery	Unspecified 'alternative surgical programme'			✓	
Totals				5	6	1

Source: Knight (2007c).

Table 10.2 Non-veterinary student outcomes: humane teaching methods compared with harmful animal use

Study	Discipline	Humane option	Total students (humane option)	Humane method superior	Equivalent learning outcomes	Humane method inferior
1 Cohen & Block 1991	Psychology	Field study (feral pigeons)			✓	
2 Clark 1987	Physiology	Computer simulation			✓	
3 Cross & Cross 2004	Biology (high school)	Computer simulation	74 (38)			✓
4 Dewhurst et al. 1988	Physiology	Computer simulation			✓	
5 Dewhurst & Meehan 1993	Physiology, pharmacology	Computer simulations	65		✓	
6 Dewhurst et al. 1994	Physiology	Computer simulation	14 (6)		✓	
7 Downie & Meadows 1995	Biology (undergraduate)	Models (rats)	2,913 (308)		✓	
8 Fowler & Brosius 1968	Biology (high school)	Video	156	✓		
9 Henman & Leach 1983	Pharmacology	Biovideograph videotape recordings	50	✓		
10 Hughes 2001	Pharmacology	Computer simulations			✓	
11 Kinzie et al. 1993	Biology (high school)	Interactive videodisc	61		✓	
12 Leathard & Dewhurst 1995	Physiology (medicine)	Computer simulation	156		✓	

13	Leonard 1992	Biology (undergraduate)	Interactive videodisc	142	✓		
14	Lieb 1985	Biology (high school)	Lecture			✓	
15	Matthews 1998	Biology (undergraduate)	Computer simulation	20 (12)			✓
16	McCollum 1987	Biology (high school)	Lecture	350 (175)	✓		
17	More & Ralph 1992	Biology (undergraduate)	Computer courseware	184 (92)	✓		
18	Phelps *et al.* 1992	Physiology (nursing)	Interactive videodisc		✓		
19	Samsel *et al.* 1994	Physiology (medicine)	Computer simulations	110	✓		
20	Strauss & Kinzie 1994	Biology (high school)	Interactive videodisc	34 (17)		✓	
21	Velle & Hal 2004	Biology (high school)	Computer simulation	64	✓		
	Totals				8	11	2

Source: Knight (2007c).

Table 10.3 Additional benefits of humane teaching methods in veterinary education

	Study	Veterinary discipline	Humane option	Benefits of humane option (other than decreased harmful animal use)
1	Allen & Chambers 1997	Surgery	Computerised tutorial	Increased surgical skill
2	Baillie *et al.* 2003	Clinical skills (bovine)	Virtual reality simulator	Customisation of learning experience, repeatability, superior skill acquisition and development
3	Baillie *et al.* 2005a	Clinical skills (bovine)	Virtual reality simulator	Customisation of learning experience, repeatability, superior skill acquisition and development
4	Baillie *et al.* 2005b	Clinical skills (bovine)	Virtual reality simulator	Customisation of learning experience, repeatability, superior skill acquisition and development
5	Buchanan *et al.* 2005	Biochemistry	3D animations	Superior understanding of complex biological processes
6	Dhein & Memon 2003	Continuing education	Internet-based curriculum	Overcomes obstacles of time and distance, decreased costs, facilitates lifelong learning
7	Dyson 2003	Anaesthesia	CD-ROM	Improved anaesthetic knowledge
8	Ellaway *et al.* 2005	Unspecified	Virtual learning environment	Increased flexibility of use
9	Erickson & Clegg 1993	Physiology	Computer simulations	Greatest student satisfaction
10	Galle & Bubna-Littitz 1983	Clinical skills (canine)	Cadaver	Repeatability
11	Greenfield *et al.* 1994	Surgery	Models	Decreased student and faculty objections to harmful animal use
12	Hawkins *et al.* 2003	Clinical skills (small animal)	Video	Improved diagnostic skills
13	Hines *et al.* 2005	Pathology (systemic)	Virtual learning environment	Greater understanding, student satisfaction, increased flexibility of use
14	Holmberg *et al.* 1993	Surgery	Model	Decreased student stress, repeatability
15	Howe & Slater 1997	Surgery	Sterilisation programme	Improved surgical and anaesthetic skills, including atraumatic tissue handling, increased understanding of the pet overpopulation problem and the role of the veterinarian in combating it, increased awareness of the activities of humane organisations

16	Howe et al. 2005	Surgery	CD-ROM	Increased practice of techniques, enhanced preparedness for laboratories, greater student satisfaction
17	Josephson & Moore 2006	Anatomy	DVD	Customisation of learning experience to individual needs, possibly improved examination results
18	Kumar et al. 2001	Anatomy	Ethically sourced cadavers	Compliance with animal use regulations, elimination of student and faculty objections to the use of purpose-killed animals, integration of clinical perspectives and ethics early in the curriculum
19	Linton et al. 2005	Anatomy	Computer simulation	Rapid access to related views such as radiographs, increased learning efficiency and student confidence
20	Modell et al. 2002	Anaesthesia	Human patient simulator	Realism, improved confidence coping with complex clinical problems, better examination results
21	Mori et al. 2006	Surgery	Model	Repeatability, improved surgical skill
22	Pinkney et al. 2001	Parasitology	Computer tutorial	Improved examination scores
23	Richardson et al. 1994	Surgery	Sterilisation programme	Improved surgical and anaesthetic skills, including atraumatic tissue handling, increased understanding of the pet overpopulation problem and the role of the veterinarian in combating it, increased awareness of the activities of humane organisations
24	Rudas et al. 1993	Unspecified	*Hypermedia*	Increased teaching efficiency, decreased cost
25	Silva et al. 2003	Surgery	Cadavers	Improved surgical skill
26	Simpson & Meuten 1992	Clinical skills	Pathology specimens	Repeatability
27	Smeak et al. 1991	Surgery	Haemostasis model	Improved surgical skill acquisition
28	Waldhalm & Bushby 1996	Unspecified	Personal computer	Enhanced information retrieval and communication, improved student attitudes towards computers, increased employer perception of computer literacy
29	Whithear et al. 1994	Microbiology	Hypermedia database	Greater autonomy and more active learning, facilitation of postgraduate learning

Source: Knight (2007c).

Veterinary surgical training

Humane surgical teaching methods compared with traditional harmful animal use have included models or surgical simulators (Johnson & Farmer 1989, Greenfield *et al.* 1994 & 1995, Smeak *et al.* 1994, Olsen *et al.* 1996, Griffon *et al.* 2000) and cadavers (Carpenter *et al.* 1991, Bauer *et al.* 1992, Pavletic *et al.* 1994).

Skills assessed in surgical laboratories included psychomotor (all), ligation (Olsen *et al.* 1996, Griffon *et al.* 2000), intestinal *anastomoses* (conjoining of two segments, normally after resection of a diseased portion), and *coeliotomy* closures (closure of an incision through the abdominal wall; Carpenter *et al.* 1991), *gastrotomy* closures (closure of an incision into the stomach; Smeak *et al.* 1994), and *ovariohysterectomies* (removal of the uterus and ovaries; Griffon *et al.* 2000).

Overall, the surgical skills instilled by these humane alternatives were at least equivalent to those achieved via traditional harmful animal use. In fact, three surgical studies demonstrated superior surgical skills when humane alternatives were used. Johnson and Farmer (1989) found that inanimate models were superior to live animals in teaching basic psychomotor skills. Olsen and colleagues (1996) demonstrated that a fluid *haemostasis* model (to allow practice of haemorrhage control) was at least as effective as a live dog *splenectomy* (removal of the spleen) for teaching blood vessel ligation and division. In fact, students using the model completed their ligatures more quickly, with fewer errors. They successfully tied more square knots, their ligatures were tighter, and their instrument grip was superior. The initial scepticism of these students regarding the use of properly designed inanimate models for teaching these surgical skills altered dramatically. Griffon and colleagues (2000) found that 20 veterinary surgical students trained using plastic surgical simulators performed ovariohysterectomies on live dogs with greater skill than 20 classmates trained via cadavers. In all cases the ability to use the models repeatedly contributed to the superior surgical skills of the students who used them.

Five studies demonstrated equivalent surgical skills when humane alternatives were compared with harmful animal use (Carpenter *et al.* 1991, Bauer *et al.* 1992, White *et al.* 1992, Pavletic *et al.* 1994, Greenfield *et al.* 1994 & 1995). Carpenter and colleagues (1991) and Bauer and colleagues (1992) demonstrated equivalent surgical skill acquisition using cadavers as the humane option, and Greenfield and colleagues (1994 & 1995) demonstrated a similar result using soft tissue organ models. White and colleagues (1992) found that veterinary students from an alternative surgical laboratory programme had surgical skills equivalent to those of students with a standard laboratory experience, albeit after some initial hesitancy among the alternative students during their first live animal surgery.

One study demonstrated inferior surgical skill acquisition using the humane option. Smeak and colleagues (1994) compared live animal gastrotomy skills of two groups of 20 students, one of which had practised the procedure using a hollow organ model, and the other of which had practised using a live animal. While the authors found no significant difference in overall gastrotomy closure technique, the students performing the procedure for a second time on a live animal were significantly quicker. Anaesthetic time is an important surgical consideration; hence this was considered a superior learning outcome. However, the plastic model used in this study was deficient, being more fragile and stiffer than living gastric tissue, with suture pull-through occurring despite appropriate technique and tension – even though the model was found to be effective for teaching instrument use, needle placement, atraumatic tissue handling, and tissue inversion.

Learning outcomes were compared both in the short term (Johnson & Farmer 1989, Carpenter *et al.* 1991, Bauer *et al.* 1992, White *et al.* 1992, Smeak *et al.* 1994, Greenfield *et al.* 1994 & 1995, Olsen *et al.* 1996, Griffon *et al.* 2000) and in the long term. As stated previously, Pavletic and colleagues (1994) studied new graduates from the Tufts University veterinary class of 1990, which included 12 students who had participated in an alternative small animal medical and surgical procedures course. This course involved the use of ethically sourced cadavers and additional clinical rotations in the small animal surgery (four weeks), small animal medicine (one week), and intensive care (one week) units. These students and 36 of their conventionally trained peers were assessed by questionnaires sent to their employers, who were asked to rate their competence at the time of hiring and 12 months later. There was no significant difference on either occasion in the abilities of the conventional and alternative graduates when performing common surgical, medical, and diagnostic procedures, in their attitudes towards performing orthopaedic or soft tissue surgery, in their confidence in performing the specified procedures, or in their ability to perform them unassisted.

The success of humane surgical training has also been reported for UK veterinary graduates. The UK is the only major region of the developed world where harmful animal use has been removed from the veterinary surgical curriculum for decades. Instead students gain practical experience by assisting with beneficial surgeries during extramural rotations at private veterinary clinics. In 1998 Fitzpatrick and Mellor (2003) surveyed graduates from all veterinary schools in Great Britain and Ireland who had graduated within the previous five years. Ninety-five per cent of respondents were working full time in veterinary practice. Graduates rated extramural studies as 'very useful' for three subjects, two of which were small animal surgery and cattle surgery.

Non-surgical veterinary disciplines

Historically and contemporarily, surgery and physiology respectively have been the disciplines that have resulted in the greatest harmful animal use during veterinary education. Disciplines other than surgery were poorly represented in comparative studies of veterinary student performance, totalling only two studies.

Abutarbush and colleagues (2006) found that a CD-ROM was more effective than a live animal demonstration by an instructor of the correct method for inserting a nasogastric tube into a horse. Students using the CD-ROM performed significantly better on a test of knowledge, were more confident, and were significantly quicker at successfully inserting a nasogastric tube into a live horse than their traditionally instructed peers.

Fawver and colleagues (1990) found that first-year veterinary students learnt cardiovascular physiology principles more efficiently from interactive videodisc simulations than from live animal laboratories, resulting in both student and staff time savings.

Related non-veterinary disciplines

Fourteen studies examined learning outcomes of university biology, medical, nursing, pharmacology, physiology, and psychology students. A very slightly higher proportion of non-veterinary students than veterinary students achieved superior or equivalent learning outcomes using humane alternatives (92.9 per cent or 13/14 versus 90.9 per cent or 10/11).

Cardiovascular physiology students achieved equivalent learning outcomes using computer simulations (Clarke 1987, Dewhurst *et al.* 1988), and superior learning outcomes using an interactive video programme (nursing students, Phelps *et al.* 1992), compared with animal-based laboratories. They also rated computer simulations as superior for learning (medical students, Samsel *et al.* 1994). Intestinal physiology students working independently with a computer program gained equivalent knowledge, for one-fifth of the cost, to students using freshly killed rats (Dewhurst *et al.* 1994, Leathard & Dewhurst 1995). Physiology and pharmacology students using computer simulations performed as well as students using traditional animal laboratories (Dewhurst & Meehan 1993). Pharmacology students achieved superior learning outcomes using biovideograph videotapes (Henman & Leach 1983), and equivalent learning outcomes overall (superior initially in each of five experiments, but possibly with inferior long-term recall of experimental details) using computer simulations (Hughes 2001), to outcomes achieved via animal-based laboratories. Biology students achieved superior (computer simulations, More & Ralph 1992) or equivalent (videodisc, Leonard 1992; models, Downie & Meadows 1995) learning outcomes using alternatives to dissections. In addition, the videodisc group used only half the time required by the traditional laboratory group.

Only one study of non-veterinary students demonstrated inferior learning outcomes when the humane teaching option was used. Eight undergraduate biology students who dissected foetal pigs scored significantly higher on an oral test with *prosected* foetal pigs than 12 students who studied using a computer simulation ('MacPig', Matthews 1998). However, MacPig is considered insufficiently detailed for college-level biology instruction (Balcombe 1998).

Effect of chronology on comparative studies

Of the 11 studies comparing veterinary student learning outcomes, 8 were published prior to 1996 and hence were more than a decade old at the time of this survey. Of the 21 papers describing non-veterinary student learning outcomes, 18 were more than a decade old. Hence, a considerable number of these studies examined humane teaching methods such as films, interactive videodiscs, and early computer simulations, which have largely been superseded by more advanced alternatives, particularly in the field of computer simulations. The laboratories these alternatives were designed to replace, such as animal dissections, demonstration experiments, or surgical exercises, have, on the other hand, remained largely unaltered. It is a damning indictment of harmful animal use that even such relatively antiquated alternatives resulted in superior or equivalent learning outcomes in almost all cases. It is likely that comparative studies of modern alternative teaching methods would yield an even higher proportion of studies demonstrating superior learning outcomes to harmful animal use.

Additional advantages of humane teaching methods

Veterinary disciplines

Twenty-nine papers describing humane teaching methods in veterinary education that did not involve comparisons with harmful animal use (although comparison with non-harmful teaching methods sometimes occurred) illustrated other advantages of these methods (Table 10.3). These included:

- increased flexibility of use (Dhein & Memon 2003, Ellaway *et al.* 2005, Hines *et al.* 2005), including increased capacity to customise learning exercises (e.g. to work at one's own pace, explore areas of deficient understanding, or repeat exercises; Galle & Bubna-Littitz 1983, Simpson & Meuten 1992, Holmberg *et al.* 1993, Whithear *et al.* 1994, Baillie *et al.* 2003, 2005a & 2005b, Dhein & Memon 2003, Howe *et al.* 2005, Josephson & Moore 2006);
- improved acquisition and development of clinical skills (Baillie *et al.* 2003, 2005a & 2005b, Hawkins *et al.* 2003), surgical skills (Smeak *et al.* 1991, Richardson *et al.* 1994, Allen & Chambers 1997, Howe & Slater 1997, Silva *et al.* 2004, Mori *et al.* 2006), and anaesthetic skills (Richardson *et al.* 1994, Howe & Slater 1997, Dyson 2003);
- superior understanding of systemic pathology (Hines *et al.* 2005) and complex biological processes (specifically, the spatial relationships of, and interactions between, intracellular molecules; Buchanan *et al.* 2005);
- rapid access to relevant anatomical views such as radiographs, with increased learning efficiency (Linton *et al.* 2005);
- enhanced preparedness for laboratories (Howe *et al.* 2005) and even, on occasion, a more realistic laboratory experience (Modell *et al.* 2002);
- improved examination results in parasitology (Pinkney *et al.* 2001), anaesthesiology (Modell *et al.* 2002), and anatomy (Josephson & Moore 2006);
- decreased student stress (Holmberg *et al.* 1993) and increased student satisfaction (Erickson & Clegg 1993, Hines *et al.* 2005, Howe *et al.* 2005) and confidence (Linton *et al.* 2005), including when coping with complex clinical problems (Modell *et al.* 2002);
- enhanced student information retrieval and communication abilities, improved student attitudes towards computers, and increased employer perception of computer literacy (Waldhalm & Bushby 1996);

- facilitation of ongoing undergraduate and postgraduate learning (Whithear *et al.* 1994, Dhein & Memon 2003);
- increased teaching efficiency and decreased costs (Rudas 1993, Dhein & Memon 2003);
- increased compliance with animal use regulations, elimination of student and faculty objections to the use of purpose-killed animals, and early integration of clinical perspectives and ethics into curricula (Greenfield *et al.* 1994, Kumar *et al.* 2001);
- increased understanding of the pet overpopulation problem and the role of the veterinarian in combating it, and increased awareness of the activities of humane organisations, when veterinary students participate in animal shelter sterilisation programmes (Richardson *et al.* 1994, Howe & Slater 1997).

Unusually, one alternative teaching model, the Bovine Rectal Palpation Simulator, was described in three of these papers (Baillie *et al.* 2003, 2005a & 2005b). Bovine rectal palpation is most commonly conducted for the purposes of pregnancy diagnosis. Designed to teach the necessary skills via a *haptic* system, this model applies anatomically appropriate tension to a student's fingers, depending on their spatial location inside a simulated cow.

Baillie and colleagues found that students using the simulator were able to customise their learning experiences according to individual need, and that they performed better when examining real cows for the first time than their traditionally trained peers. However, bovine rectal palpation is not normally harmful or unduly stressful unless performed repeatedly. Hence, this animal use was not considered harmful for the purposes of this review, although some repeated use does occur in veterinary practical classes.

Related non-veterinary disciplines

Numerous papers describing related non-veterinary disciplines that did not involve comparisons with harmful animal use (although comparison with non-harmful teaching methods was sometimes made) have also illustrated additional advantages, and occasionally disadvantages, of humane teaching methods. More than 500 such papers published from 1974 to 2006 were identified during this review. Many of these described the development, validation, and effects on surgical planning, skill levels, and other surgical or educational outcomes of endoscopic or other surgical simulators. The validation of such a simulator assesses its ability accurately to predict real surgical skill levels, and is typically achieved when experienced and inexperienced surgeons demonstrate differing skill levels while using the simulator. A rigorous analysis of these papers is well beyond the scope of this review. However, examples of papers of particular interest to biomedical educators include:

- twenty-three papers demonstrating increased development of endoscopic diagnostic and surgical skills, or conventional surgical skills, particularly in suturing, achieved by medical students or practitioners through the use of computer-based virtual reality or haptic, endoscopic, or other surgical simulators;
- three papers indicating equivalent learning outcomes when alternatives to the dissection of human cadavers were used. These were prosected specimens, a stereoscopic slide-based auto-instructional programme, and interactive videodiscs and computer simulations, and the students were medical students (Prentice *et al.* 1977, Jones *et al.* 1978) and human gross anatomy, pre-nursing and allied medical profession students (Guy & Frisby 1992);
- one paper describing the use of the *pulsatile organ perfusion trainer*, in which arteries in waste organs (commonly from slaughterhouses, although ethically sourced cadavers could also be used) are perfused with an artificial blood solution via a pulsatile pump, for training in both minimally invasive and conventional surgical techniques. Unlike many surgical simulators, this model allows practice of haemostatic techniques. Even complex operations, such as colorectal and anti-reflux procedures, may be performed;
- two papers discussing the potential for globalised surgical teaching via *telesurgery* (the introduction of miniaturised cameras into patients during surgery);
- one paper describing the development of a realistic intubation simulator for practising *endotracheal intubation* (insertion of a tube into the trachea of an anaesthetised patient, for the administration of anaesthetic gas/oxygen mixtures) in dogs. Another study demonstrated that the human intubation skills of paramedical students who were trained using a simulator were equivalent to those of students trained on human subjects;
- a study by Huang and Aloi (1991) which demonstrated the improved learning outcomes of undergraduate biology students who used computer simulations of dissections. Similarly, Holt and colleagues (2001) demonstrated that computer-aided learning (CAL) can be effective in teaching endocrinology to medical students;

- one study demonstrating increased satisfaction and examination results for cardiovascular physiology students when computer simulations were used (Lilienfield & Broering 1994). Another demonstrated cardiovascular physiology knowledge acquisition equivalent to that gained from a textbook, although these medical students rated the computer simulation superior for reinforcement and review (Specht 1988);
- a study by Dewhurst and Jenkinson (1995) demonstrating that computer simulations generally saved teaching staff time, were less expensive, and were an effective and enjoyable mode of undergraduate biomedical student learning.

Additional references are provided by Knight (2007c).

A small minority of studies demonstrated inferior learning outcomes when humane teaching methods were used. For example, Rogers and colleagues (1998) demonstrated lesser basic surgical skill acquisition (the ability to tie a square knot correctly) by medical students when a CAL program was used instead of a lecture and feedback seminar. Student comments suggested that the lack of feedback in this CAL model resulted in the significant difference between these two learning outcomes. Caversaccio and colleagues (2003) found that a virtual simulator enhanced understanding of endonasal surgery, but failed to make an impact on operating room performance. The simulator's effectiveness was limited by the absence of force feedback, subtle handling of the joysticks, and considerable time consumption. Gerson and Van Dam (2003) found that medical residents trained to perform a *sigmoidoscopy* (allowing endoscopic examination of the interior of the sigmoid colon) via traditional bedside teaching techniques achieved greater skill than those trained using an endoscopy simulator. Furthermore, a review of 30 randomised controlled trials assessing any training technique using at least some elements of surgical simulation found that none of the methods of simulated training (computer simulation, models, cadavers) was conclusively superior to another, or to standard surgical training, primarily of medical students and practitioners (Sutherland *et al.* 2006). These studies emphasise the importance of ensuring that humane teaching methods are well designed, and focused on achieving the specific learning outcomes desired.

Animal welfare and regulatory benefits

Advantages of humane alternatives other than educational efficacy include the saving of substantial numbers of animal lives. Few countries record the numbers of animals used for educational purposes, and, of those that do, most consider only live vertebrate use, and fail to include invertebrates, or vertebrates killed for dissections. In addition, the small proportion of non-harmful use is rarely, if ever, differentiated from overall animal use. Consequently, estimates of the numbers of animals harmed for educational purposes are problematic.

Nevertheless, it is clear that those numbers are substantial. On the basis of the most recent governmental figures available, 207,500 animals were used for education and training in the EU in 2008, or 1.7 per cent of the 12.0 million total, of which the overwhelming majority were mice, rats, and poikilotherms (EC 2010b).

The US is the largest national user of laboratory animals (Table 2.1). Balcombe (2000b) estimated that approximately 9 million vertebrates and a similar number of invertebrates were used in biomedical education in the US in 2000.

Australia is the fourth-largest national user of laboratory animals (Table 2.1). In those four of Australia's eight states and territories supplying figures for 2008, 1.0 million animals, or 19.0 per cent (after elimination of rounding errors) of the 5.1 million official total, were used for educational purposes (HRA 2010).

Canada is the sixth-largest national user of laboratory animals (Table 2.1). Around 98,600 animals, or 4.3 per cent of the 2.3 million official total used by Canadian participants in the CCAC programme in 2008, were used for teaching purposes. While overall recorded animal use has risen significantly in the past two decades in Canada, the use of animals for teaching purposes has fluctuated fairly steadily between 2.8 per cent and 5.7 per cent (CCAC 2009).

Apart from directly saving large numbers of animal lives and positively impacting animal welfare, humane teaching methods facilitate increased compliance with legislative and code of practice requirements restricting educational or other scientific animal use, which exist in an increasing number of countries (Balcombe 2000a). In Australia, for example, the *Australian Code of Practice for the Care and Use of Animals for Scientific Purposes*, which is enforceable via the animal welfare legislation of each state and territory, requires alternatives to the use of animals for educational and other scientific purposes, wherever possible (NHMRC 2004).

The considerable importance of factors such as these is expected to increase further as society becomes ever more conscious of animal welfare concerns (Siegford *et al.* 2005), and consequently less willing to permit harmful animal use for educational purposes (Scalese & Issenberg 2005).

Summary

Sufficient studies have been conducted to draw certain conclusions about the efficacy of humane teaching methods in imparting surgical skills or knowledge. Well-designed humane alternatives generally perform at least as well as methods that rely on harmful animal use, in some cases achieving superior learning outcomes. These outcomes have included superior surgical, anaesthetic, and other clinical skill acquisition and development, superior understanding of complex biological processes, increased learning efficiency, and improved examination results.

In addition, increased teaching efficiency and decreased costs, along with enhanced potential for customisation and repeatability of the learning exercise, frequently result. Increased student confidence and satisfaction, enhanced preparedness for laboratories, and decreased student stress may also occur, as may enhanced student information retrieval and communication abilities, improved student attitudes towards computers, and increased employer perception of computer literacy.

Increased compliance with animal use legislation or regulations, elimination of student and faculty objections to the use of purpose-killed animals, and early integration of clinical perspectives and ethics into veterinary curricula may also result. Substantial numbers of animal lives are saved, and some evidence suggests that veterinarians trained without harmful animal use may develop higher animal welfare standards (see Chapter 11, pp. 171–4), potentially benefiting their future patients. They may also gain increased understanding of the pet overpopulation problem and the role of the veterinarian in combating it.

Rather than continue to rely on harmful animal use, the evidence clearly indicates that veterinary and other biomedical educators can best serve their students and animals, and can minimise financial and time burdens on their faculties, by introducing well-designed humane teaching methodologies.

Detailed information about the alternatives available for various academic disciplines is provided by Jukes and Chiuia (2003) and on websites such as www.vetmed.ucdavis.edu/Animal_Alternatives/main.htm and www.clive.ed.ac.uk. Synopses of surgical simulators designed for medical students and practitioners are provided at www.virtualsurgery. vision.ee.ethz.ch.

Links to libraries from which a variety of alternatives may be borrowed, along with free online computer simulations, comprehensive alternatives databases, academic reviews of leading alternatives, and hundreds of educational studies of alternatives organised by discipline, are also available through the websites www.HumaneLearning.info and www.EURCA.org (Knight 2007c).

11
Effects of Harmful Animal Use on Students

The adverse impacts of invasive animal use in biomedical education are not limited to the animals used. Adverse impacts may also be experienced by students. These may include long-term health impacts, impacts on learning, and impacts on the development of attitudes towards animal welfare.

Such adverse attitudinal impacts can profoundly alter the ability of veterinarians to safeguard and promote good welfare for their patients and for animals generally, contrary to both societal expectations and professional obligations stipulated in veterinary codes of professional conduct (e.g. RCVS 2010). In addition, veterinary students work with animals more than most other biomedical students, and justifications proposed for invasive animal use are stronger in the case of veterinary education than for any other educational discipline. Accordingly, when we seek to ascertain the effects of invasive educational animal use on the development of attitudes towards animals and their welfare, it is once again highly instructive to examine veterinary students and veterinarians closely.

Health hazards

The highly toxic chemicals used to preserve anatomy specimens between dissections present health hazards, raising the potential for legal and financial liability should students suffer exposure-related adverse health effects. Formaldehyde, for example, is widely used. It is very efficiently absorbed from inspired air, and is a potent genotoxin, primarily via DNA–protein cross-linkage. Epidemiological studies have associated occupational formaldehyde exposure with increased cancer risks at various sites, including the nasal cavities, nasopharynx, lung, brain, pancreas, prostate, colon, and *lymphohaematopoietic* (bone marrow, splenic, blood, and lymphatic) system (Orsière *et al.* 2006). Accordingly, the use of formaldehyde is banned from several applications under the EU's *Biocidal Products Directive* (98/8/EC). Exposure levels in gross anatomy laboratories used by Japanese medical students, for example, may significantly exceed permitted safe limits (Takayanagi *et al.* 2007, Uchiyama 2010). In my experience and that of veterinary student colleagues between 1998 and 2006, recommended safety guidelines such as the use of gloves, gowns, and masks are not commonly met with full compliance in veterinary schools (see Knight 2007c for details). These veterinary schools all had high standards, and this limited survey suggests that there may be a wider problem internationally, rather than a problem with these specific schools.

Impacts on learning

For many students, participation in harmful live animal use, as occurs in physiology demonstration or surgical training laboratories, generates powerful emotional experiences and a high degree of stress, which have great potential to affect learning adversely. To a lesser degree, this is also true of anatomy dissection laboratories.

In 1999, 370 veterinary students were surveyed at the University of Illinois College of Veterinary Medicine about the learning value of the first-year terminal physiology labs they had participated in (Stull 2000). The majority of students were critical of the laboratories:

> For each lab, ONE person in FOUR got to place an IV catheter; most of us were too preoccupied with having to kill the dog that physiology wasn't concentrated on; there was nothing surgical about the procedure.

> It was difficult to get any great understanding of physiology because we worried most of the time about not having our dog bleed to death or die of anesthetic overdose before the experiment was over. In the end, what I learned about physiology (cardio and resp) I taught myself from the notes.

> We watched videos on everything that was done. I learned just as much from that. The lab I did participate in turned out to be a bloody mess. I was more worried about keeping the animal alive than I was about what was going on physiologically from the drugs we administered. I also had a problem using animals where in the end we euthanized (when we could benefit from watching videos).

Nothing that was covered in those labs could not have been learned from a demo, or a video. The guilt I felt for participating outweighed all beneficial aspects of the experience.

A minority of students considered the experience beneficial, however:

> These laboratories are the only time we get to work on a living breathing animal our first two years of vet school. When you get to junior year and you work on animals that are expected to live, you wish you had had more experience with this kind of thing. I admit the physiology I got out of it was minimal but the experience of working with an animal that bleeds was very helpful.

> I learned a tremendous amount from the labs and really tried to make the most of them. However, I think most of that experience would've been gained later on in 3rd year with surgeries, as I think I learned more about surgeries, catheters, and that type of thing than about physiology, although I did learn some. I think clinical experience would be a better use of resources to gain more experience with blood draws, catheter insertions, animal handling and that sort of thing that we really didn't get a whole lot of here.

Following this survey and adverse publicity resulting from a student campaign, all these laboratories were cancelled in 2000, saving the lives of more than 100 pigs, dogs, rats, and rabbits annually. An Animal Use Policy was also passed, formally allowing conscientious objection to harmful animal use in the veterinary school, and in 2003 a similar policy was passed covering the entire university.

Similarly, in 2001, third-, fourth-, and fifth-year veterinary students at New Zealand's Massey University were surveyed to ascertain their views about the learning value of their third-year terminal physiology laboratories (Anon. 2001). Although students in junior years were generally supportive of the labs, support was much lower among those completing their final year. By their fifth year, a majority of the 105 respondents felt that neither the physiological knowledge, surgical knowledge, nor skill gained justified the use of live animals, and that alternative learning methods could provide the required knowledge of physiological principles. They also expressed their preference for learning live tissue handling (i.e. surgical skills) in a more clinically oriented veterinary subject.

The fifth-year students were completing their final, clinical year of the veterinary course. The key difference between these students and the junior students surveyed is that the former would have had a greater understanding of the clinical roles they would shortly be expected to fulfil as new graduates, and consequently a more realistic understanding of the value of these terminal physiology laboratories in imparting useful skill and knowledge.

Prior to this survey 68 sheep were killed annually in these six terminal laboratories. Once again, following this survey and a student campaign, two of the six labs were cancelled in 2002, and the other four were changed to demonstration laboratories, in which a total of eight sheep were killed annually. The Physiology Department planned to end those labs as well within two years.

Unfortunately, many faculties have not been amenable to student requests for humane teaching methods, or have allowed them only after protracted struggles – as occurred at both the University of Illinois and Massey University. Unsurprisingly, faculty opposition to strong student desires for humane teaching methods frequently results in conflict. Adverse effects can include costs in time and money, and damage to relationships between students and faculty of sufficient severity that marked adverse impacts on student learning, student expulsions, and lawsuits against universities may all result. However, the strength of the case in favour of humane teaching methods, and the potential for adverse publicity generated by such conflicts, are such that universities usually – but not always – rapidly yield to student requests for humane teaching methods once lawsuits are initiated or publicity is generated.

At the University of Colorado School of Medicine in 1992, for example, Dr Safia Rubaii – who was then a medical student – was initially failed for refusing to participate in physiology demonstration laboratories in which dogs and frogs were killed. She sued her university for US$95,000 in damages and costs in 1995, and the university was also required to introduce a humane learning option (McCaffrey 1995). These terminal laboratories have since been completely eliminated from the course.

A substantial number of countries have completely banned the harmful use of animals in school (usually) or university education. In a smaller group, including England, Germany, India, Italy, the Netherlands, and the US, the rights of students to educational methods that do not violate their conscientiously held ethical or religious beliefs against harming animals are protected by various constitutional safeguards, legislation, policies, or conventions, which have contributed to several successful lawsuits by students (see Knight 2007c for details; see also Francione & Charlton 1992, Balcombe 2000a & 2000b).

Attitudinal impacts

Although often overlooked, the attitudinal impacts of harmful animal use in education may be profoundly important for both students and staff.

Veterinary attitudes towards animal welfare

To investigate veterinary attitudes towards animal welfare, in 2005 a colleague and I reviewed the animal welfare policies of the World Veterinary Association, the American Veterinary Medical Association (AVMA), the British Veterinary Association, the Australian Veterinary Association (AVA), and the New Zealand Veterinary Association (NZVA) (De Boo and Knight 2006).

More specifically, we reviewed the positions of each of these associations on five animal use practices that have raised considerable animal welfare concerns among both scientists and the general public:

- small 'battery' cages for laying hens;
- gestation crates for pregnant sows;
- small crates and nutritionally deficient diets for veal calves;
- the use of animals in scientific research and education;
- the tail docking of dogs.

We compared the results of our veterinary review with research data on public attitudes towards these five animal use practices. There was widespread and persistent public concern about each of these practices in all the countries surveyed. In contrast, many of these specific concerns were not addressed clearly in the five veterinary associations' positions. The associations either lacked positions on, or were not categorically opposed to, the close confinement of laying hens, pregnant sows, and veal calves – although the NZVA did recommend time limits on the use of sow gestation crates, and both the NZVA and the AVA recommended group, rather than individual, housing of veal calves.

The only practice to which the public and the associations appeared to share opposition was the cosmetic tail docking of dogs, although the AVMA did not take a firm stance against this. In the case of animal experimentation, both the general public and the veterinary profession appeared to support experimentation for human medical research to some degree, although public opinion remained very critical. The results clearly suggested that veterinarians lag behind the general public in their desire for animal welfare reforms.

To scrutinise the attitudes of veterinarians towards animal welfare more closely, in 2005 I conducted a detailed review of the positions of the AVMA (Knight 2008f). With more than 72,000 veterinary members working in private and corporate practice, government, industry, academia, and the uniformed services by 2005 – the largest membership of any veterinary association worldwide – and claiming to act as 'a collective voice for its membership and for the profession' (AVMA 2005), the AVMA was ideally suited to this purpose.

The results indicated that although the AVMA did not support all practices commonly resulting in poor animal welfare, it clearly supported a range of practices that do, in some cases contrary to both substantial scientific evidence and public opinion. Farm animal husbandry practices supported by the AVMA included the 'forced' moulting (primarily through nutritional restriction and manipulation of light/dark cycles), unanaesthetised beak trimming, and close confinement of laying hens. These practices jointly constitute the greatest violations of welfare inflicted on large number of animals worldwide, and have been condemned on animal welfare grounds by a succession of official bodies on the basis of overwhelming scientific evidence (Knight 2008f). At the time of writing they were still largely supported by the AVMA, albeit with certain provisos (AVMA 2011).

The ongoing support of leading national and international veterinary associations for such practices, contrary to both rigorous animal welfare science and strong public opinion, is clearly disturbing. It raises the question: what factors in veterinary selection, education, or practice result in such poor attitudes towards animal welfare?

Education of veterinary students

The importance of educating veterinary students about animal welfare issues, and of assisting their development of the critical thinking skills needed to negotiate these controversies successfully, is increasingly recognised, and the consequent incorporation of bioethics and critical reasoning courses into veterinary curricula worldwide is ongoing. Nevertheless, the proportion of veterinary students receiving such formal education remains small.

On the other hand, there is a 'hidden curriculum' endorsing harmful animal use which remains commonplace in veterinary schools worldwide. Students are typically required to harm and kill animals in preclinical subjects such as anatomy (dissection, often of purpose-killed animals or animals from ethically debatable sources), physiology, biochemistry, and pharmacology (demonstration experiments on living animals – usually of long-established scientific concepts, with animals usually killed during or after the experiment). Students are typically required to practise surgical and anaesthetic skills via anaesthetising healthy animals, conducting surgical procedures on them, and killing any survivors at the end (not all survive these frequently prolonged operations).

Hence, the majority of veterinary students receive minimal or no formal education in animal welfare issues or critical reasoning, and are directly required by some of their professors to harm and kill animals during their education. These professors are seen by most students as the leaders of the veterinary profession, and the unwritten message delivered is that harming and killing healthy animals is not only condoned, but is required, in order to become a veterinarian, and, further, that animal welfare concerns are subservient to human interests of debatable merit.

Desensitisation-related phenomena

Unsurprisingly, therefore, the decreasing awareness among veterinary students of animal sentience (specifically, the hunger, pain, fear, and boredom of dogs, cats, and cows) over the duration of their veterinary courses (Paul & Podberscek 2000, Levine *et al.* 2005); the decreased likelihood of fourth-year students providing analgesia compared with second- or third-year students (Hellyer *et al.* 1999); and the inhibition of the normal development of moral reasoning ability during the four years of veterinary school (Self *et al.* 1991 & 1996) have all been documented in veterinary journals. Along with inadequate curricular attention to animal welfare science, the human–animal bond, and the development of critical reasoning ability and ethics (Self *et al.* 1994, Williams *et al.* 1999), the harmful use of animals during veterinary education is a likely cause of such phenomena (Serpell 2005, De Boo & Knight 2005 & 2006). These desensitisation-related phenomena appear to be psychological adaptations that enable previously caring students to withstand what could otherwise be intolerable psychological stresses resulting from requirements to harm and kill sentient animals in the absence of overwhelming necessity.

Consequently, the replacement of harmful animal use with humane teaching methods is likely to result in veterinarians with more positive attitudes towards animal welfare, which is likely to directly benefit their animal patients.

Summary

Invasive animal use in biomedical education may adversely affect students and staff in several ways. Anatomy specimens are preserved between dissections using highly toxic chemicals. These create significant health hazards, and the subsequent potential for legal and financial liability should students suffer exposure-related effects. In my experience and that of colleagues from leading veterinary schools, recommended safety guidelines such as the use of gloves, gowns, and masks are often met with only partial compliance.

For many students, participation in harmful live animal use, such as that occurring in physiology demonstration or surgical training laboratories, generates powerful emotional experiences and high levels of stress, which have great potential to adversely affect cognitive processes such as learning. Veterinary student accounts indicate that they are commonly distracted from relevant scientific concepts by the plight of their animals, and the necessity of concentrating on maintaining life and anaesthetic depth for the duration of the experiment. A minority of students, however, consider such experiences beneficial for learning.

Over time, student participation in harmful animal use appears to contribute to a range of desensitisation-related phenomena which adversely impact awareness of animal welfare problems and the desire to take appropriate action to redress them. Such adverse attitudinal impacts have profound potential to decrease the ability of veterinarians to safeguard and promote good welfare for their patients and for animals generally. This is contrary to both societal expectations and professional obligations stipulated in veterinary codes of professional conduct (e.g. RCVS 2010). Comparison of the positions of leading national and international veterinary associations on controversial animal use issues with data from public surveys clearly indicates that veterinarians lag behind the general public in their desire for animal welfare reforms. In some cases regressive veterinary association positions are also contrary to substantial evidence from the field of animal welfare science.

Accordingly, the replacement of harmful animal use in education with humane teaching methods is likely to result in veterinarians with more positive attitudes towards animal welfare. This is likely to directly benefit their animal patients. However, faculty opposition to student desires for humane teaching methods has been common internationally, frequently resulting in conflict. Adverse effects have included considerable time and financial costs, damage to relationships between students and faculty sufficient to markedly affect learning, student expulsions, lawsuits against universities, and significant adverse publicity.

Part V

Conclusions and Policy Recommendations

12
The Costs and Benefits of Animal Experimentation

Unanswered questions about the precise psychological abilities of laboratory animals inevitably result in a degree of uncertainty about the nature and magnitude of the suffering likely to result from invasive procedures and protocols. It has been theorised that only those species with brain structures such as a cerebral cortex and thalamus and the capacity for synaptic feedback between the two have the capacity for sentience (Butler 2000). If true, this would include most vertebrates, with the possible exception of *chondrichthyes* (sharks and rays) and *cyclostomes* (lampreys and hagfish), but not invertebrates. Although many invertebrates display complex behaviour, there is little evidence of brain structures comparable to those believed to support consciousness in higher animals (Nicol 2010). Cephalopods, however, display electroencephalogram patterns (i.e. electrical activity or 'brainwaves') similar to those associated with differing wake–sleep states of consciousness in vertebrates (Edelman & Seth 2009), demonstrating their possible sentience, and also the limits of our understanding about what neuroanatomical mechanisms are necessary for sentience, and about which animals possess them, and to what degree.

Where such doubt exists, it seems reasonable to apply a precautionary principle – to assume until proven otherwise that suffering *may* occur, and to consider restrictions on procedures likely to cause such suffering. Similar precautionary principles are, after all, enshrined in fundamental social institutions, because they are considered to be rational, reasonable, and humane. The Western legal system, for example, generally assumes innocence until guilt is proven beyond reasonable doubt. Where such doubt remains, judicial punishment should be withheld.

Conversely, a precautionary principle could be applied in favour of human patients or consumers who may potentially benefit from animal experimentation: where healthcare advances or other human utility *may* result, perhaps such experiments should proceed until lack of benefit is proven beyond reasonable doubt.

These viewpoints represent diametrically opposite positions, consistent with ideologies that consider the interests of animals or people, respectively, to be overwhelmingly more important than the interests of the other group. These are but two of a diverse range of philosophical, cultural, and religious viewpoints on our moral duties towards animals and people that could be applied (Armstrong & Botzler 2003, Council of Europe 2006, Busche 2008).

It is the opinion of various philosophers (e.g. Singer 1990, Busche 2008) that a reasonable and rational balance should be sought between upholding the interests of people and those of laboratory animals. This requires balanced consideration of the interests of both groups: primarily, the human interests in obtaining benefits such as new clinical interventions and the identification of toxins, and the interests of animals in avoiding harms such as involuntary confinement, social disruption, various forms of suffering, and death. This *utilitarian* assessment aims to achieve the 'greatest good for the greatest number', and considers the interests of all those affected, whether human or other creatures likely to be capable of experiencing states as 'good' or as less desirable.

Such a utilitarian assessment forms the basis for most regulation governing scientific animal use. *Directive 2010/63/EU on the protection of animals used for scientific purposes*, which directs such animal use in all EU member states, asserts that it is essential, on both moral and scientific grounds, to ensure that the scientific or educational validity, usefulness, and relevance of each use of an animal are carefully assessed. It specifically requires that the likely harms to the animal should be balanced against the expected benefits of the project (EU 2010).

Fortunately, although uncertainties remain, sufficient evidence exists to draw some key conclusions about the likely costs to animals, and benefits to humans, of animal experimentation overall.

Animal costs

Numbers and species of animals used

Accurate estimation of global laboratory animal numbers is markedly impeded by the lack of published national statistics. Among those countries that do publish data, lack of standardisation hinders interpretation. Nevertheless, using a validated prediction model based on animal study publication rates, Taylor and colleagues (2008) were able to provide evidence-based estimates of national and global laboratory animal use (Chapter 2). Application of the most accurate statistical methods revealed that, worldwide in 2005, at least 126.9 million non-human vertebrates were subjected to fundamental or medically applied biomedical research, toxicity testing, or educational use; were killed for the provision of experimental tissues or as surplus to requirements; or were used to maintain established GM strains (Knight 2008a, Taylor *et al.* 2008). Although this total represents the most accurate recent, evidence-based estimate of global laboratory animal use, it remains highly conservative, due to several factors. It also excludes certain categories that raise ethical concerns, such as the use of advanced foetal developmental stages and of certain invertebrate species believed to possess significant capacity for suffering.

On the basis of the latest EU reports (EC 2010a & 2010b; see also Chapter 3), the overwhelming majority of animals used are mice (around 59 per cent) and rats (around 18 per cent). Poikilotherms (fish, amphibians, and reptiles; 10 per cent) and birds (6 per cent) are the next-largest groups.

Clearly, almost all animals subjected to scientific procedures are higher vertebrates. These animals possess the neuroanatomical mechanisms and psychological capacities necessary to experience significant pain, fear, and psychological distress. Great apes in particular have advanced emotional, psychological, and social characteristics which enhance their potential for suffering in laboratory environments. To a lesser degree capacities for sentience may also exist in the small proportion of other animals, such as certain invertebrates, which are used.

Effects of stressors on welfare and scientific outcomes

A wide variety of stressors have the potential to cause significant stress, fear, and possibly distress in laboratory animals. These stressors may be associated with the capture of wild-sourced species such as primates to supply laboratories or breeding centres; with transportation, which may be prolonged for some animals; with laboratory housing and environments; and with both routine and invasive laboratory procedures (Chapter 4).

A large minority of all procedures are markedly invasive. These include procedures resulting in death (whether or not the animals were conscious), surgical procedures (excluding minor operative procedures), major physiological challenges, and the production of GM animals. On the basis of Canadian figures, the proportion of markedly invasive procedures has ranged between approximately 29 per cent and 44 per cent over the past decade, with a figure of 40 per cent recorded in 2008. The prevalence of such procedures may be increasing (CCAC 2009; see also Chapter 3). A sizeable majority of all procedures utilise no anaesthetics of any kind. Canadian figures are not available, but in Britain procedures without anaesthesia have fluctuated between approximately 59 per cent and 69 per cent of recorded totals over the past two decades, with a figure of 67 per cent recorded in 2009 (Home Office 2010; see also Chapter 3).

To assess animal impacts further it would be helpful to know the frequency of analgesic use, the degree of correlation between markedly invasive procedures and anaesthetic or analgesic use, and the prevalence of environmental enrichment and socialisation opportunities. Unfortunately, such information is largely lacking.

Nevertheless, a large number of studies have demonstrated that the stress caused by laboratory housing and environments, and by routine laboratory procedures, may result in profound, statistically significant distortions in a range of physiological indices, including cardiovascular parameters and serum concentrations of glucose and various hormones. Behaviour may be markedly altered, and behavioural stereotypies and increased aggression may develop over time, as may alterations in certain neuroanatomical parameters and even cognitive capacities (Balcombe *et al.* 2004, Balcombe 2006, Baldwin & Bekoff 2007; see also Chapter 4). Some of these effects are also likely to be sequelae of other stressors such as invasive procedures and transportation.

Unsurprisingly, the chronic stress experienced by most laboratory animals may result in immunocompromisation, and subsequently increased susceptibility to a range of pathologies. As well as creating significant animal welfare and ethical problems, such stressors and their effects on laboratory animals may distort a wide range of experimental outcomes, such as those dependent on accurate determination of physiological, behavioural, or cognitive characteristics.

Human benefits

Human clinical and toxicological utility

In the EU, around 42 per cent of all procedures in 2008 were focused on the development, production, or safety evaluation of clinical interventions and other products, almost all of which were intended for human use (EC 2010a).

The historical and contemporary paradigm that animals are reasonably predictive of human outcomes provides the basis for such widespread use in toxicity testing and biomedical research aimed at combating human diseases. However, their use persists for historical and cultural reasons, rather than because it has been demonstrated to be scientifically valid.

In fact, most systematic reviews published in peer-reviewed scientific journals have demonstrated that animals are insufficiently predictive of human outcomes to provide substantial benefits during the development of human clinical interventions or the assessment of human toxicity. In only 2 of 20 such reviews located during a comprehensive survey (Knight 2007a & 2008b) did the authors conclude that animal models were either significantly useful in contributing to the development of clinical interventions or substantially consistent with clinical outcomes. Furthermore, one of these conclusions was contentious.

Included were reviews examining the human clinical utility of invasive chimpanzee experiments, of highly cited animal experiments published in leading scientific journals, and of experiments approved by ethics committees at least partly on the basis of specific claims that these animal experiments were likely to lead to concrete advances in human healthcare. Seven additional reviews also failed to demonstrate reliable predictivity of human toxicities such as carcinogenicity and teratogenicity. Results in animal models were frequently equivocal or inconsistent with human outcomes (Chapters 5–6).

Those systematic reviews investigating the human clinical utility of chimpanzee experimentation or the toxicological utility of carcinogenicity bioassays are particularly significant, given that other animal models are even less likely to be generally predictive of human outcomes than chimpanzees, and other fields of toxicity testing are even less likely to provide public health benefits than carcinogenicity testing. By extrapolation, our current reliance on animal models of humans must be questioned in all fields of clinically oriented biomedical research and toxicity testing.

Causes of poor human utility

The likely causes of the poor human clinical and toxicological utility of animal models include inherent genotypic and phenotypic differences between human and non-human species, the distortion of experimental outcomes arising from stressful experimental environments and protocols, and the poor methodological quality of many animal experiments, which is apparent from numerous systematic reviews of experimental utility (Knight 2007a & 2008b). No reviews were identified in which a majority of animal studies were of good methodological quality.

Problems arising from stressors and poor methodological quality might theoretically be minimised, although fundamental changes to the practice of laboratory animal science would be required, given their widespread prevalence. However, limitations resulting from interspecies differences are likely to be technically and theoretically impossible to overcome.

Non-animal alternatives

Much animal use in biomedical research and toxicity testing can be replaced by a broad range of non-animal methodologies, either individually or in combination (Chapter 8). Prior to commencing any biomedical research project or toxicological evaluation, researchers should collate and examine all relevant existing data to determine which, if any, remaining studies are necessary. Current commercially motivated restriction of access to important proprietary data has significant implications for the detection of toxicity in compounds under development, as well as for the wider development of clinical interventions, animal welfare, and resource allocation. New regulatory mechanisms are therefore necessary to enhance the sharing and assessment of such data. In other cases, information is publicly available, but its assessment is inadequate.

During pharmaceutical development or toxicity assessment, qualities such as absorption, distribution, *in vivo* concentrations in various body compartments, organ systems affected, toxicity, efficacy, clearance, and metabolic fate can be predicted to varying degrees through physicochemical evaluation, chemical grouping with interpolation or extrapolation of properties, and computerised modelling, including the use of structure–activity relationships and expert systems.

A variety of tissue cultures are available, including immortalised cell lines, embryonic and adult stem cells, and organotypic cultures. *In vitro* assays using bacterial, yeast, protozoal, mammalian, or human cell cultures can predict a wide range of toxic and other endpoints. To increase the spectrum of toxins detected, individual assays may be combined as test batteries. Human hepatocyte cultures and metabolic activation systems may allow identification of metabolic pathways and of metabolites produced, and assessment of organ–organ interactions. Toxicity onset, magnitude, and levels over time can be measured by analysis of biomarkers in the outflow of perfused cultures. Microarray technology will increasingly allow genetic expression profiling of toxins, increasing the speed of detection, well before more invasive endpoints.

Nevertheless, to predict the likely effects of test compounds most accurately, human studies will remain necessary. The safety for volunteers and predictivity for diverse patient populations of human clinical trials may be increased through measures such as microdosing, staggered dosing, larger study populations, and longer durations. Additional human-based studies could use surrogate tissues, advanced imaging modalities, and human epidemiological, sociological, and psychological investigations to shed light on illness aetiology and pathogenesis, and to facilitate the development of safe and effective clinical interventions.

Weighing the costs and benefits

When considering costs and benefits overall, one cannot reasonably conclude that the benefits accruing to human patients or consumers, or to those motivated by scientific curiosity or profit, exceed the costs incurred by animals subjected to scientific procedures. On the contrary, the evidence indicates that actual human benefit is rarely – if ever – sufficient to justify such costs. The more speculative the human benefit becomes – as in the case of fundamental (rather than clinically applied) biological research, for example, which constitutes a major category of laboratory animal use – the more obvious the resultant disparity.

It is, in fact, only possible to conclude that such research is ethically justified if a profoundly unequal weighting is applied in which relatively minor or infrequent human benefits are considered more important than the significant adverse impacts commonly experienced by laboratory animals. However, this position is increasingly inconsistent with our growing understanding of the psychological and social characteristics of laboratory animals, including their ability to experience suffering and pleasure; of the impacts on laboratory animals resulting from laboratory environments and procedures; and of the moral obligations that stem from this knowledge. It also profoundly distorts the utilitarian cost/benefit analysis that fundamentally underpins most policy and regulation governing animal experimentation.

Even when animal interests are marginalised in such ways, it remains far from clear that animal experimentation is justifiable. The famous 'greatest happiness' definition of utilitarianism formulated by the noted utilitarian philosopher Mill (1971) asserted that one should always act so as to produce the greatest happiness for the greatest number of individuals. However, animal experimentation is unreliably – and frequently poorly – predictive of human outcomes, and consumes enormous financial resources and human expertise, which are then unavailable to other research fields. Those potentially affected include patients, consumers, and scientists. The moral implications are profound when consumers suffer serious toxic reactions to products assessed as safe in animal studies (Chapter 5), or if patients with serious conditions are denied effective clinical interventions partly because potentially more efficacious research fields are under-resourced.

The costs of scientific animal use can be surprisingly wide-ranging, as illustrated by considering the case of veterinary students. Students are exposed to highly toxic chemicals used to preserve anatomy specimens between dissections, at levels that may exceed recommended safe limits. Their learning may be adversely affected by the stresses arising from the conflict created by simultaneous requirements to inflict grievous harm and to care for their animal subjects during laboratory classes. Finally, studies have demonstrated that senior veterinary students may suffer cognitive phenomena such as decreased awareness of animal sentience and inhibition of normal development of moral reasoning ability. They are also less likely to provide analgesia, and it appears likely that the animal welfare standards of veterinarians are adversely affected as a result (Chapter 11).

Such desensitisation-related phenomena are almost certainly adaptive coping mechanisms that enable previously caring students to withstand substantial psychological stress resulting from curricular requirements to harm and kill animals. The implications for laboratory animal scientists and technicians subjected to similar stresses of even greater frequency and magnitude, and the potential for adverse impacts on the welfare of animals in their care, are both obvious and disturbing.

The case for invasive animal research is further weakened by consideration of the potential offered by the diverse and growing array of non-animal or non-harmful research, testing, and educational methodologies. In the latter case the evidence is remarkably consistent: students using humane alternatives designed to impart knowledge or surgical skills virtually always achieve learning outcomes at least equivalent – and in some cases superior – to those achieved through traditional, harmful animal use.

Non-animal research and testing methodologies are not yet able to answer all potential questions about humans, particularly given current technological limitations. Yet the same criticism applies to animal models, which have a considerably more limited capacity for further development.

On the other hand, non-animal models can offer certain important advantages. Particularly when humans or human tissues are used, such methods may generate faster, cheaper results which are more reliably predictive for humans, and may yield greater insights into human biochemical processes. Such logistical considerations are increasingly important, given the unprecedented challenges posed by high-throughput US and EU toxicity testing programmes such as HPV and REACH respectively, as well as rising social pressures to find alternatives to laboratory animal use. These pressures are increasingly resulting in legislative or regulatory changes, such as the seventh amendment to the European *Cosmetics Directive 76/768/EEC*.

The case of chimpanzees

Few research issues generate as much controversy as invasive chimpanzee experimentation. The unequalled phylogenetic proximity of chimpanzees to humans makes them potentially superior to all other laboratory species for use as human models in toxicity experiments and in pathological or therapeutic investigations it would be hazardous to conduct on humans. However, their use also raises greater animal welfare and bioethical concerns than the use of virtually any other laboratory species, because of their advanced emotional, psychological, and social characteristics. These characteristics markedly increase their ability to suffer when chimpanzees are born into unnatural captive environments, or captured from the wild (as many research chimpanzees once were), and then subjected to confinement, social disruption, and involuntary participation in potentially harmful biomedical research.

The justifications proposed for invasive chimpanzee experimentation rely on the important contributions advocates claim such research has made to the advancement of biomedical knowledge, and, in particular, to combating major human diseases. However, a recent large-scale citation analysis of the medical utility of chimpanzee experimentation indicated that the benefits are significantly less than is sometimes claimed. Half of the randomly selected published chimpanzee studies examined were not cited by any subsequent papers, apparently generating data of questionable value which made little obvious contribution to the advancement of biomedical knowledge. In addition, closer examination failed to identify any chimpanzee study that made an essential contribution, or, in a clear majority of cases, a significant contribution of any kind, to papers describing methods efficacious in combating human diseases (Knight 2007b & 2008c; see also Chapter 5). These conclusions have since been confirmed by additional studies examining the contribution of chimpanzee research to AIDS research and vaccine development (Bailey 2008), cancer research (Bailey 2009), and HCV research and vaccine development (Bailey 2010, Bettauer 2010).

The costs to chimpanzees enrolled in such experiments include involuntary confinement in laboratory settings, social disruption, and participation in potentially harmful research protocols. Recent studies have established beyond all reasonable doubt that the effects of laboratory confinement and procedures, especially long term, can be severe. Many captive great apes, including chimpanzees recently retired from US laboratories (Bradshaw *et al.* 2008), show gross behavioural abnormalities such as stereotypies, self-mutilation or other self-injurious behaviour, inappropriate aggression, fear, or withdrawal (Brüne *et al.* 2006, Bourgeois *et al.* 2007). It is increasingly acknowledged that such abnormal behaviours resemble symptoms associated with human psychiatric disorders such as depression, anxiety disorders, eating disorders, and post-traumatic stress disorder, and that pharmacological treatment modalities similar to those applied to human patients may be appropriate, and, indeed, morally compelled, for severely disturbed animal patients (Brüne *et al.* 2006, Bourgeois *et al.* 2007). Long-term therapeutic combination with positive reinforcement training, environmental enrichment, and social and environmental modification may be necessary in severe cases (Bourgeois *et al.* 2007).

Consideration of an analogous legal scenario is illuminating. Although these highly sentient creatures are innocent of causing any human grievance, including the serious diseases we attempt to induce in them, we sometimes subject chimpanzees to conditions that would cause widespread outrage if used to punish the most heinous of human criminals – for years on end, and, in some cases, for decades. Bradshaw and colleagues (2008) observed: 'The costs of laboratory-caused trauma are immeasurable in their life-long psychological impact on, and consequent suffering of, chimpanzees.' In contrast, human criminals are not normally punished until proven guilty beyond reasonable doubt. The application of such differing treatment standards to humans and chimpanzees reveals a lack of 'humanity' paradoxically less characteristic of chimpanzees, than of ourselves.

The logic of Bradshaw and colleagues' corollary is elementary, yet compelling: 'In human traumatology, the first step in treatment is to arrest its causes. This implies that prevention and treatment of chimpanzee psychopathology entails considering the factors and institutions that have brought chimpanzees to the point of irreversible distress: in simple terms, desisting from using apes as biomedical subjects in lieu of humans is compelled if trauma is not to be perpetuated.'

The remarkable biological characteristics of chimpanzees (which are rare in their own right) and their advanced emotional, psychological, and social characteristics (which have some similarities to those of humans) create a strong ethical basis for acknowledging the necessity of respecting at least the most basic and essential interests of chimpanzees, such as their interests in avoiding death, pain, suffering, and captivity (Cavalieri & Singer 1993, Morton 2000). When reasonable consideration is afforded to the interests of both humans and chimpanzees, it cannot be concluded that invasive chimpanzee experimentation is ethically justifiable.

Implications

Ethical oversight

Almost all the chimpanzee experiments examined in Chapter 5 would have been approved by at least one institutional ethics committee obliged to permit only those experiments likely to result in substantial benefits, given the considerable animal welfare, bioethical, and financial costs inherent in chimpanzee experimentation. To varying degrees, similar costs are incurred when all higher vertebrates, and probably some invertebrates, are subjected to invasive experimentation.

Although the concept of ethical review is sound, these results demonstrate that its implementation is currently flawed. This flaw appears to have resulted from an over-reliance on the assumption that invasive experiments on chimpanzees and other laboratory animals were likely to be of substantial use in advancing biomedical knowledge. The approval of large numbers of such experiments despite their questionable merits clearly demonstrates a widespread failure of ethical oversight, adding significantly to previous concerns about the effectiveness of ethics committees in safeguarding laboratory animal subjects (Schuppli & Fraser 2005).

By approving these experiments on the basis of unfounded assumptions about their likely benefits, the ethics committees responsible failed in their duty to society, and to the animals they were charged with protecting.

Model validation

Despite substantial ongoing progress in the development of non-animal alternatives, compliance with the spirit of the 3Rs and, indeed, with the letter of some associated regulations remains poor in many sectors of government, academia, and industry. Continued reliance on animal models is understandable when they are truly required by regulators for the licensing of drugs and chemicals. The position of such regulators themselves appears less justifiable, however. Some apparently 'feel more comfortable' with animal data (O'Connor 1997), or even believe animal tests are inherently valid simply because they are conducted in animals (Balls 2004), despite a large and growing body of evidence to the contrary (Chapters 5–6). Animal-based toxicology, at least, appears to be 'frozen in time, using and accepting the same old animal models again and again, often without stringent examination of their validity' (Hartung 2008a; see also Leist *et al.* 2008b).

It is clear that rather than relying on assumptions of human utility, we should subject animal experimental models to the same standards of scientific scrutiny currently required for non-animal alternatives prior to regulatory acceptance. Such scientific *validation* has traditionally involved the demonstration in multiple independent laboratories that the test in question is relevant to and reliable for its specified purpose (*practical validation*; Balls *et al.* 1995), such as the prediction of a certain *in vivo* outcome. However, it is not always scientifically necessary, or even logistically possible, to conduct multi-centre practical studies. Hence *weight-of-evidence validation*, also known as *validation by retrospective analysis* (ICCVAM 1997, OECD 2003), may be conducted by assessing existing data in a structured, systematic, and transparent manner – provided that data of sufficient quantity and quality are available (Balls & Combes 2006). Where practical validation studies do occur, they should adhere to best practice standards designed to ensure good methodological quality, including, for example, statistical justification of sample sizes, randomised allocation to test groups, and blinded treatment and assessment of results. Where feasible, inter-laboratory reproducibility should be demonstrated (Balls & Combes 2005).

Scientific validation should lead to a reasoned overall assessment that sufficient evidence exists to demonstrate that a model is, or is not, relevant to and reliable for the specified purpose, or that insufficient evidence exists to be reasonably certain either way. In some cases, an interim assessment may be made until further evidence becomes available (Balls & Combes 2006).

The European Centre for the Validation of Alternative Methods was created by the EC in 1991 to fulfil the requirements of European *Directive 86/609/EEC on the protection of animals used for experimental and other scientific purposes.* This Directive required the EC and its member states actively to support the development, validation, and acceptance of methods which could replace, refine, or reduce the use of laboratory animals (ECVAM 2010a). The US equivalent is the Interagency Coordinating Committee on the Validation of Alternative Methods, which has similar goals. Despite the high standards required for successful validation, between 1998 and 2010, 30 distinct tests or categories of test methods that could replace, reduce, or refine laboratory animal use were assessed and declared by ECVAM to be scientifically valid, of which 24 achieved regulatory acceptance (ECVAM 2010b).

Unlike non-animal models, animal models are generally assumed to be reasonably predictive of human outcomes in preclinical drug development, toxicity testing, and other fields of biomedical research without the need to undergo formal validation studies. Yet the 27 systematic reviews examined in Chapters 5–6 demonstrate the invalidity of assuming that animal models are reliably predictive of human outcomes, even when in use for long periods, without subjecting them to critical assessment.

Clearly, formal validation should be consistently applied to all proposed experimental models, regardless of their animal, non-animal, historical, contemporary, or possible future status. Model choices should also be based on mechanistic relevance to the hypothesis under investigation; on specificity, sensitivity, predictivity, durations, and other relevant scientific data; and on ethical, legal, and resource considerations. Such standards should be upheld regardless of historical or contemporary acceptance. Models not meeting these standards should be discarded, whether animal-based or otherwise.

Likely benefits of adherence to such standards would include greater selection of models truly predictive for human outcomes, increased safety of people exposed to chemicals that have passed toxicity tests, increased efficiency during the development of human pharmaceuticals and other therapeutic interventions, and decreased wastage of human and financial resources and animal lives.

Reduction and refinement

Where scientific animal use does continue, a range of strategies should be implemented to reduce animal numbers and minimise their suffering. Strategies to decrease animal numbers can be applied at the level of individual experiments (*intra-experimental reduction*). Improvements in experimental design and statistical analysis are key examples. Reduction strategies can also focus on implementing best practice policies at institutions where animal experiments occur (*supra-experimental reduction*). Improved education and training of staff, the inclusion of statisticians on animal ethics committees, and retrospective evaluation of experiments all fall into this category. More distantly related developments such as international harmonisation of testing requirements also have great potential to reduce animal numbers (*extra-experimental reduction*). *Refinement strategies* aim to minimise animal suffering. They include appropriate use of analgesic and anaesthetic agents (which remain underutilised), non-invasive imaging modalities, telemetric devices, positive reinforcement techniques, and environmental enrichment and socialisation opportunities. In many cases simultaneous implementation of different 3Rs strategies can have synergistic effects, improving both animal welfare and scientific quality.

Ethically justifiable research

Animal research ranges from field studies of free-living (wild) populations, through non-invasive behavioural or psychological studies of sanctuary or laboratory populations, to mildly harmful invasive experimentation, more harmful experimentation, and, finally, protocols resulting in major harm or death. According due respect to animal interests does not require the termination of all animal-based research. Bioethical concerns are minimised in non-invasive observational, behavioural, or psychological studies of free-living or sanctuary populations.

For animals with advanced psychological and social characteristics, such as chimpanzees, there are risks of boredom and associated pathology in sanctuary settings, unless these settings are highly enriched. Offering chimpanzees the choice to participate in behavioural or psychological studies may, in fact, constitute a valuable form of environmental enrichment (Matsuzawa *et al.* 2006). When participation remains truly voluntary – rather than coerced through conditional fulfilment of important needs such as food, water, or social contact with compatible conspecifics – bioethical concerns are minimised. Such studies are permissible under existing bans on great ape experimentation in countries such as Sweden, and are consistent with the US *Chimp Haven is Home Act* (2007), which prohibits further research on chimpanzees retired to federal sanctuaries, other than non-invasive behavioural studies (Participatory Politics & Sunlight Foundations 2008; see also 'Retirement of laboratory chimpanzees' in Chapter 13).

Limiting animal experimentation to non-invasive observational, behavioural, or psychological studies of free-living or sanctuary populations would inevitably restrict the range of scientific questions that could be investigated. It would, however, strike the correct ethical balance between satisfying the interests of animals and satisfying those of human beings.

13

Regulatory Developments and Policy Recommendations

Existing regulations

Increasing social concern about invasive animal use in biomedical research and toxicity testing has driven the evolution of legislation and related regulations in many countries and regions. Such regulations seek to restrict laboratory animal use to only those instances where non-animal alternatives are considered scientifically inadequate, for investigating questions of sufficient importance to justify animal use. Institutes are typically required to establish animal ethics committees (IACUCs in the US) to ensure compliance with such regulations.

The scope of such regulations is typically limited to living vertebrates. However, some variations exist. The US – the largest national user of laboratory animals (Table 2.1) – excludes mice, rats, birds, fish, reptiles, and amphibians from protection under the *Animal Welfare Act*, even though mice and rats make up the overwhelming majority of experimental subjects. And Great Britain – the seventh-largest national user – protects one species of octopus, *Octopus vulgaris* under its *Animals (Scientific Procedures) Act* (1986). *Directive 2010/63/EU on the protection of animals used for scientific purposes*, which directs such animal use in all EU member states, similarly protects all living cephalopods (EU 2010).

Certain categories of animal use resulting in ethical concerns are commonly excluded from national regulations. Examples include animals bred for laboratory use but found to be surplus to requirements, the killing of animals used to supply organs or other tissues, and the use of animals to maintain established GM strains. Unusually, *Directive 2010/63/EU* – the most modern regulatory example at the time of writing – includes in its scope animals intended for use as well as those actually used in scientific procedures, animals bred for tissue harvesting, and procedures resulting in both the creation and maintenance of GM animals. However, the killing of surplus animals and those used as tissue donors remains excluded from the Directive.

European Union

The EU is the only major world region publishing harmonised statistics on laboratory animal use for its member states. The *Treaty of Lisbon*, which regulates the governance and organisation of the modern EU, recognises animals as sentient beings, thereby providing an important legal and philosophical basis for animal protection regulations (EU 2007).

In 1986, the Council of Europe opened for signature by member states European Treaty Series No. 123 (ETS 123), the *European Convention for the Protection of Vertebrate Animals used for Experimental and Other Scientific Purposes*. It requires the use of non-animal methods 'if reasonably and practicably available', or justification of animal use otherwise, in which case numbers must be minimised, along with pain, suffering, or lasting harm (Council of Europe 1986). The Convention has the legal status of an international treaty.

Similarly, the 1986 European *Directive 86/609/EEC on the protection of animals used for experimental and other scientific purposes* stated in Article 7.2: 'An experiment shall not be performed if another scientifically satisfactory method of obtaining the result sought, not entailing the use of an animal, is reasonably and practicably available' (Kulpa-Eddy 2006). Compliance with such Directives is mandatory for EU member states. However, the Directive did not require ethical review or compulsory authorisation of animal experiments.

In 2010 this Directive was superseded by *Directive 2010/63/EU on the protection of animals used for scientific purposes* (EU 2010). The new Directive strengthened the protection afforded to laboratory animals, explicitly requiring systematic ethical evaluation and authorisation of scientific protocols. It states in its preamble that:

> It is also essential, both on moral and scientific grounds, to ensure that each use of an animal is carefully evaluated as to the scientific or educational validity, usefulness and relevance of the expected result of that use. The likely harm to the animal should be balanced against the expected benefits of the project. Therefore, an impartial project evaluation independent of those involved in the study should be carried out as part of the authorisation process of projects involving the use of live animals.

The scope of the Directive was also broadened. Protection was extended from living vertebrates to include foetal forms of mammals in the final third of their gestation, independently feeding larval forms, living cephalopods, animals bred specifically so their organs or tissues can be harvested for scientific purposes (although their killing is explicitly excluded), those used in the creation and maintenance of GM strains, and those *intended* for use in scientific procedures, in addition to those actually used. The use of non-human primates is restricted – particularly in the case of great apes – although not fully prohibited (see 'Regulatory developments' below).

Implementation of the 3Rs is emphasised:

> When choosing methods, the principles of replacement, reduction and refinement should be implemented through a strict hierarchy of the requirement to use alternative methods. (preamble)

> Member States shall ensure that, wherever possible, a scientifically satisfactory method or testing strategy, not entailing the use of live animals, shall be used instead of a procedure. (Article 4.1)

> Member States shall ensure that the number of animals used in projects is reduced to a minimum without compromising the objectives of the project. (Article 4.2)

> Member States shall ensure refinement of breeding, accommodation and care, and of methods used in procedures, eliminating or reducing to the minimum any possible pain, suffering, distress or lasting harm to the animals. (Article 4.3)

Notably, the new Directive specifies an upper limit of pain, suffering, and distress above which animal use is not normally permissible. Procedures resulting in severe pain, suffering, or distress which is likely to be long-lasting and incapable of being ameliorated are permissible only in exceptional circumstances.

United States

As stated, the US is the largest national user of laboratory animals. The first national US legislation governing laboratory animal use was *Public Law 89–544* (the *Animal Welfare Act*) of 1966, which was amended by *Public Law 99–198* (the *Improved Standards for Laboratory Animals Act*) of 1985 (AWIC 2011). Subsequent regulations have clarified intentions and expectations for publicly funded research to ensure compliance with the law. Several important clarifications are provided in the USDA's *Animal Care Policy Manual*. The first key expectation is that potentially painful or distressing procedures will be recognised as such. The USDA's Animal Care Policy 11 defines such procedures as 'any procedure that would reasonably be expected to cause more than slight or momentary pain or distress in a human being...'. Examples provided include procedures that are potentially painful (ocular and skin irritancy testing, terminal surgery – although this would normally be alleviated by anaesthesia) or distressing (food/water deprivation, noxious electrical shock, paralysis or immobility). Advice is also provided to aid in the recognition of pain and distress (USDA 2011a).

The second key expectation described in Policy 12 is that researchers will make sufficient, genuine efforts to seek replacement, reduction, and refinement methods. Research proposals are expected to include a rationale for using live animals, including the appropriateness of the species and animal numbers, and a description of procedures or methods designed to limit pain and discomfort to a level considered unavoidable in the conduct of 'scientifically valuable research'. A written description must be provided of the methods and sources used to consider alternatives to procedures that may cause more than momentary or slight pain or distress, with a written assurance that proposed procedures do not unnecessarily duplicate previous work. For long-term projects, such consideration of alternatives should normally be undertaken at least once every three years. However, for federally mandated animal testing (e.g. for product testing), the written narrative need only include a citation of the appropriate government agency's regulation and guidance documents (USDA 2011b).

The third key expectation is that any alternative methodologies identified will be incorporated to the appropriate degree into research protocols (Kulpa-Eddy & Adams 2008).

Additional regulatory guidance is to be found in the *Public Health Service Policy on Humane Care and Use of Laboratory Animals* and the *U.S. Government Principles for the Utilization and Care of Vertebrate Animals Used in Testing, Research, and Training*. Principle III of these Government Principles briefly refers to reduction and replacement, and Principle IV to refinement of laboratory animal use (OLAW 2002). Further guidance is provided in texts such as the *Guide for the Care and Use of Laboratory Animals* published by the Institute for Laboratory Animal Research (ILAR 2011). In addition, the Animal Welfare Information Center (AWIC) at the USDA National Agricultural Library conducts extensive literature searches on a cost-recovery basis for researchers seeking alternatives, provides information on laboratory animal care, and runs workshops on complying with the information requirements of the *AWA*.

Great apes

Bans on great ape experiments

The advanced psychological, emotional, and social capabilities of chimpanzees and other great apes have resulted in increasing concerns about subjecting them to invasive experimentation, and a growing number of national bans have subsequently been implemented through either legislative or policy provisions.

In the UK, for example, special justifications for experiments on great apes became necessary under the *Animals (Scientific Procedures) Act* (1986), and in 1997 a policy ban was placed on such experiments by the Home Office (Home Office 1998, Smith & Boyd 2002, Luy 2007). Great ape experimentation has also been banned in Sweden (regulatory restrictions since 2003, with the exception of non-invasive behavioural studies) and Austria (since 2006, unless conducted in the interests of the individual animal). Unusually, the Austrian ban also protects gibbons (Cohen 2007a, Luy 2007).

The Netherlands was the last European country to conduct invasive research on chimpanzees. It outlawed great ape experimentation from 2004 (Cohen 2007a, Luy 2007). In Italy and Norway, great apes have not been used for years, although national bans have yet to be passed. Since 1992 great apes have not been subjected to invasive research in Germany, although non-invasive cognitive and behavioural studies do occur (Kolar 2008). In 2002, the Belgian minister responsible for animal welfare announced that Belgium would be working towards a ban on all primate experiments, and a Swiss state ethics commission recently demanded that the Swiss government ban great ape experimentation (e.g. Goodman & Check 2002, Conlee *et al.* 2004, Conlee & Boysen 2005, Langley 2006, Cohen 2007a, Luy 2007). Indeed, great apes have not been used in any EU member state since at least 2002 (EC 2010a).

Japan ceased invasive research on chimpanzees in 2006 (Project R&R 2009). In Australia and New Zealand, great ape experimentation is restricted by policy (Australia; NHMRC 2003) or legislation (New Zealand, since 1999; Taylor 2001, Luy 2007), unless it is in the best interests of the individual animal or species.

Related developments continue to occur internationally. The European Principality of Liechtenstein imposed a total ban on animal experiments in 1989 (Kuhse & Singer 1998). In 2007, the Republic of

San Marino similarly banned all animal experiments (Of Human and Non-Human Animals 2007, Repubblica di San Marino 2007, Higgins 2008), and the Balearic Islands – one of the Autonomous Communities of Spain – granted basic legal rights to great apes. Due to the popularity of this development, the Spanish government was considering expanding it to include the whole of Spain (Rose 2007).

Retirement of laboratory chimpanzees

Opinion is also growing that chimpanzees should be retired at the end of their involvement in biomedical research into sanctuaries capable of providing for their social and psychological well-being for the remainder of their natural lives (e.g. Brownlee 2000, Anon. 2004a, Brent 2004). The US *Chimpanzee Health Improvement, Maintenance and Protection Act* (2000), for example, requires that chimpanzees no longer needed for biomedical research be retired to sanctuaries. The US *Chimp Haven is Home Act* (2007) repealed provisions of the *Public Health Service Act* (1946) that had permitted the removal of chimpanzees from the federal sanctuary system for research purposes. The *Chimp Haven is Home Act* prohibits the use of such chimpanzees for research, other than in non-invasive behavioural studies (Participatory Politics & Sunlight Foundations 2008).

Perhaps the first large-scale retirement of laboratory chimpanzees occurred in 2002, when Baxter Healthcare Corporation transferred more than 40 chimpanzees, and more than 80 monkeys, from its Hans Popper Primate Center (HPPC) outside Vienna, to the Home of Primates – Europe, Safaripark Gänserndorf, also in Austria. For many years the Center had used these primates to test putative vaccines for viruses such as HBV, HCV, and HIV and therapeutic plasma proteins such as Factor VIII. Both the advent of alternative testing systems and a change in research focus led to Baxter's 1998 decision to end non-human primate testing at HPPC and seek a permanent retirement site (Gonder 2004).

In 2002 the Netherlands agreed to fund the re-homing of chimpanzees by the charitable foundation Stichting AAP (Anon. 2004a). More than 100 chimpanzees – which made up the Dutch national colony – were relocated to sanctuaries, zoos, and safari parks. In 2007 similar efforts were under way in Japan (Cohen 2007a).

In the US, the New York Blood Center (NYBC) recently announced plans to retire 74 chimpanzees used in hepatitis research programmes at its Vilab laboratories in Robertsfield, Liberia. The chimpanzees will be retired to six remote African islands purchased from the Liberian government to provide a sanctuary (Leake 2007). The NYBC no longer considers such chimpanzee experiments acceptable on ethical and welfare grounds, believing that 'there are new methods for doing this kind of research' (Cohen 2007a, Times Newspapers 2007).

Regulatory developments

United States

One fundamental failure of US regulation is that alternatives need only be considered. The federal *AWA* does not actually require their use, even when they have been scientifically validated, and contains provisions specifically prohibiting the federal government from regulating the design of research or experimentation (Kulpa-Eddy & Adams 2008). Accordingly, certain states have passed legislation to prevent animal testing when validated alternatives exist. California was the first, in 2000, and was followed by New Jersey and New York, in 2007 and 2008 respectively (State of California 2000, State of New Jersey 2007, State of New York 2008).

In addition, a *Petition to the US Food and Drug Administration for [the] Mandatory Use of Non-Animal Methods* was filed in 2007 by a coalition of non-profit organisations and individual scientists seeking changes in FDA regulations. These changes would require the use of data from scientifically satisfactory non-animal test methods during FDA assessments, in lieu of corresponding animal-based methods, whenever such alternatives are available (MAP Coalition 2007, Pippin 2008).

Non-human primates

In 2007, 433 Members of the European Parliament (MEPs) signed Parliamentary *Written Declaration 40/2007* calling for urgent action to end the use of great apes and wild-caught monkeys in experiments, and for the establishment of a timetable for the cessation of all European primate experiments. The number of signatories was the highest recorded for any Declaration on an animal protection issue, and the third highest for a Declaration of any kind, since 2000. Declarations must be formally considered by European Commission officials when drawing up applicable legislation. This Declaration called for changes to European *Directive 86/609/EEC on the protection of animals used for experimental and other scientific purposes*, which allowed experiments on NHPs – around 10,000 of which are subjected to experiments in Europe annually, with the greatest recorded use occurring in the UK, followed by France and Germany (Anon. 2007, EC 2010b, NAVS 2007).

UK MEP Dr Caroline Lucas stated at the time: 'The EU is currently reviewing its rules on laboratory animals, and we must use this opportunity to immediately ban the use of primates in experiments anywhere in the EU, in favour of more modern and effective alternatives like computer modelling, tissue or cell cultures and micro-dosing.' Swedish MEP Jens Holm similarly stated: 'It is time to end experiments on primates. Primates are sentient beings and are fully capable of having feelings like humans: joy, happiness or anger. Their interests must be fully taken into account, and cruelty against them must stop' (NAVS 2007).

However, *Directive 2010/63/EU on the protection of animals used for scientific purposes*, which replaced *Directive 86/609/EEC*, continues to allow the use of NHPs, although this use is limited to fields of research considered essential for the benefit of humans or the respective NHP species. This is further clarified as including basic research (except in the case of great apes), research aimed at the preservation of the respective NHP species, and work, such as xenotransplantation and the development, manufacture, and testing of pharmaceuticals and other products, which targets human medical conditions. These should normally be potentially life-threatening or debilitating, but work on lesser conditions remains permissible under one of several loopholes described in Article 55, near the end of the Directive – albeit not for great apes. Another loophole does allow the use of great apes, but only for the preservation of the species or for research targeting human conditions that are potentially life-threatening or debilitating. Member states claiming a need to conduct research under any of the Article 55 loopholes must seek EC approval.

Due to the profound stressors and injury risks associated with the capture and transportation of wild-sourced primates, the Directive also asserts the desirability of transitioning towards the use of captive-bred animals (EU 2010).

The beginning of the end for chimpanzee experimentation?

On 17 April 2008, a bi-partisan political group introduced *The Great Ape Protection Act* to the US Congress. The bill proposed to end invasive research and testing on an estimated 1,200 chimpanzees confined in US laboratories – some for more than 40 years. For approximately 600 federally owned chimpanzees, the bill would also ensure permanent retirement to sanctuaries (HSUS & NEAVS 2008). Congressman Roscoe Bartlett, who – along with others – introduced this bill, stated: 'As a scientist who worked with chimpanzees on research projects, I believe the time has come to limit invasive research on these animals and rigorously apply existing alternatives.'

In the US, laboratory chimpanzee numbers had previously soared when the NIH implemented a breeding programme in 1986 to meet the demands of researchers seeking to study the newly emergent AIDS epidemic. Following the failure of the chimpanzee model to produce clinically useful outcomes, however, in 2007 the NIH National Center for Research Resources (NCRR) made permanent a breeding moratorium temporarily implemented in 1995 (Cohen 2007a & 2007b). Numbers steadily declined thereafter, and by October 2006, 1,133 chimpanzees remained in six US primate centres (Cohen 2007b).

Finances played a large role in the NCRR decision to reduce chimpanzee numbers. With captive chimpanzees living an average of 30 years (males) to 45 years (females), the lifetime costs of supporting them are estimated at between US$300,000 and US$500,000 (Cohen 2007a). NCRR figures indicate that 650 federally funded chimpanzees will cost a total of US$325 million to support (Cohen 2007b). Although privately funded research chimpanzees remain unaffected, the NCRR decision is expected to result in a major decline in laboratory chimpanzee numbers over the next 30 years as most are retired or die.

Hence, whether through legislation or budgetary restriction, invasive chimpanzee research may be drawing to a close in the US. Furthermore, these events may herald the beginning of the end for chimpanzee experimentation internationally. Although around half a dozen other countries also conducted chimpanzee experiments two decades ago, at the time of writing the US stood almost alone. Virtually every other country except Gabon – whose status on this issue was unclear – had ceased invasive chimpanzee experimentation. Should the US and Gabon also terminate such research, it could effectively result in the first global moratorium on invasive research for any non-human species, unless conducted in the best interests of the individual or species.

Policy recommendations

Failures of policy implementation

Contrary to the intention, and in some cases the letter, of regulations governing animal experimentation in the US, the EU, and other world regions, such research does not generally provide sufficient human benefits to justify the harms inflicted on laboratory animal subjects. This conclusion is evidenced by a sizeable and remarkably consistent body of large-scale systematic reviews (Chapters 5–6).

The overwhelming majority of animals used are mice and rats, but when animals are considered individually, the greatest ethical concerns arise when NHPs are used, and particularly great apes such as chimpanzees. The highly developed psychological and social characteristics of these animals confer unique capacities for suffering in laboratory environments.

Deficiencies in the implementation of requirements to replace, reduce, and refine animal use remain marked and widespread in many world regions. In the US, for example, despite the existence of regulatory guidance and extensive support from organisations such as the AWIC, reports indicate ongoing widespread failures in the implementation of alternative methodologies. In some cases searches for alternatives have used Internet search engines, rather than appropriate scientific bibliographic databases, including those focused on alternatives, and the selection of search terms has been cursory. Other common deficiencies have included unnecessary experimental duplication, insufficient justification of animal numbers, and inadequate recognition and alleviation of pain or distress (USDA 2000, Kulpa-Eddy & Adams 2008; see also 'Scientific resistance to alternatives', Chapter 8, p. 98). Incorrect assumptions about the human utility of animal models and a dearth of interest in implementing 3Rs strategies both appear to be contributing factors.

Recommendations for policy development

A range of basic policy initiatives are warranted to address these deficiencies.

Animals protected

First, the species protected should be broadened beyond the basic inclusion of living vertebrates to protect all additional categories that raise significant ethical concerns. Some countries currently fall far short of international norms. In the US, the denial of protection to mice, rats, birds, fish, reptiles, and amphibians under the *AWA* excludes well over 90 per cent of animals used in scientific procedures (USDA 2005, Taylor *et al.* 2008).

Regulatory protection should be based on current scientific knowledge about neuroanatomical architecture; cognitive, psychological, and social characteristics; and consequent capacity for suffering in laboratory environments and protocols. Sufficient scientific evidence exists to warrant the protection of living vertebrates, including advanced larval forms and foetal developmental stages, as well as certain invertebrates, such as cephalopods. In recognition of the evolving state of scientific knowledge, where significant doubts remain about the level of development of morally relevant characteristics, such species should be afforded the benefit of the doubt until their status is clarified. Animals used to develop or maintain GM strains, bred for organ or tissue harvesting, and bred or intended for laboratory use, including those killed when surplus to requirements, should similarly be protected, with neither their use nor their killing excluded from the ethical review and regulatory control afforded to other laboratory animals.

Species and procedures associated with high welfare risks

The advanced psychological and social characteristics of great apes such as chimpanzees render it impossible in practical terms to provide laboratory environments that satisfactorily meet their minimum psychological and behavioural requirements, which include family preservation, ample opportunities for climbing, exploring, problem solving, and playing, and considerable space (Balls 1995, DeGrazia 1996, Smith & Boyd 2002). Accordingly, the use of great apes should be prohibited, and remaining primate use very carefully scrutinised, as should other categories of animal use that pose particularly high risks to animal welfare, such as terminal or surgical procedures, major physiological challenges, and the production of GM animals. In particular, animal use should be prohibited where pain, suffering, or distress is likely to be severe or long-lasting. It must be remembered that the stress caused by such procedures is also likely to substantially alter the animals' physiology and any dependant scientific outcomes.

Scrutiny of scientific animal use

Given that most large-scale systematic reviews have demonstrated minimal human clinical or toxicological benefit from invasive animal experimentation – even when expected to produce concrete advances in human healthcare (Lindl *et al.* 2005; see also Chapters 5–6) – and that educational studies have consistently demonstrated superior or equivalent learning outcomes when life and health sciences students utilise humane learning methods (Chapter 10), the likely human benefits of scientific animal use should be scrutinised far more critically than is currently the norm, and more accurately weighed against the animal, human, and financial costs incurred. The societal values attached to laboratory animal lives and the health and safety of patients and consumers are considerable, and the corresponding public interest is substantial.

To facilitate critical review, and in recognition of the legitimacy of such public interest, proposed experimental protocols should be made available for independent scientific and public scrutiny, and subjected to independent ethical review as a condition of licensing. Where valid reasons exist, researcher anonymity should be preserved. Searches for replacement, reduction, and refinement methodologies should be thorough, and where scientifically suitable alternatives are identified, they should be used, rather than merely considered, as remains the current US requirement.

Retrospective evaluation

Consistent with legitimate public interest, and to minimise unwarranted experimental duplication, study results should be made publicly available in a timely fashion. To assess the degree to which experimental objectives were successfully met and the extent to which animals suffered, and to help inform both future research strategy and further experimental licensing decisions, retrospective evaluation of experiments should be mandatory where such experiments are likely to result in significant animal harm or financial costs, or human benefits.

Minimising duplication

To minimise experimental duplication, efforts to achieve international harmonisation of toxicity testing requirements and of the regulatory acceptance of 3Rs methods should be vigorously pursued.

Compliance with 3Rs and best practice standards

Compliance with each of the 3Rs and a range of best practice standards, before and during experiments, should be mandatory to secure experimental funding, licensing, and publication of results. Such minimum standards should include those relating to animal sourcing, housing, environmental enrichment, socialisation opportunities, appropriate use of anaesthetics and analgesics, handling, non-invasive endpoints, and statistical input during experimental design. Where journal space constraints limit the description of alternative methodologies used in studies, these should be included in supplementary online databases. This would also facilitate the transfer of alternative technologies between institutions (Gruber & Hartung 2004).

To enable animal researchers and technicians to meet the necessary standards, training and continuing professional development in 3Rs methodologies and good laboratory animal practice should be compulsory.

National initiatives

It is in the public interest to ensure that research involving animals produces the maximum possible human or animal benefit, at the minimum possible cost to animals and public finances, and that research of insufficient merit is not pursued. It is therefore desirous from both scientific and policy perspectives that 3Rs strategies are implemented far more aggressively than has been the case to date. A range of national initiatives are warranted to achieve these outcomes.

Quantitative reduction targets

The implementation of 3Rs strategies would be significantly accelerated by quantitative, binding targets for reductions in national animal use.

Expert bodies and national centres for alternatives

Where significant scientific animal use occurs, expert national bodies such as the US AWIC should be established, where necessary, to provide advice on the implementation of 3Rs strategies at the level of individual experiments and more widely. Well-funded national centres should be dedicated to the ongoing development, scientific validation, and implementation of alternative methodologies. Public funds previously spent on animal tests required by regulators could be redirected into the further development and implementation of alternatives.

Conclusions

Implementation of policies such as these would not end ethically sound animal-based research or teaching. Clear-cut examples include non-invasive observational or behavioural studies of domesticated species, or of non-domesticated species in sanctuaries or the wild; the education of veterinary students via participation in beneficial clinical or surgical procedures on genuine animal patients; and experimental treatment of animal patients genuinely suffering from severe, naturally occurring disease or injury, when conventional treatment is ineffective.

However, rigorous implementation of 3Rs methodologies would improve not only animal welfare, but also scientific standards and the rigour of subsequent results. When safe and effective human clinical interventions and reliable toxicity predictions are sought, alternative strategies may provide superior human predictivity, and greater insights into human biochemical processes, for a small proportion of the time and cost of many traditional animal assays.

Greater commitment to the development and use of alternative methodologies is clearly necessary to maximise efficiency when attempting to meet the needs of high-throughput toxicity testing programmes. It is necessary to support the ongoing development of human clinical interventions, and to meet important emerging testing needs, such as the toxicity testing of nanoparticles, genetically altered or irradiated food, and mobile phone radiation – none of which can be adequately assessed using classical animal models (Leist *et al.* 2008b).

Rigorous implementation of policies such as these would restore to animal research the balance between human and animal interests expected by society, intended by legislation, and demanded by detailed ethical review.

Glossary

3Rs – *replacement* of animal use with non-animal alternatives wherever possible in scientific procedures; *reduction* of animal numbers to the minimum possible; and *refinement* of animal use to avoid or minimise animal pain, distress, or other adverse effects suffered at any time during the animals' lives, and to enhance well-being (Buchanan-Smith *et al.* 2005). The *relative replacement* of live animal use with cells, tissues, or organs sourced from animals is considered less ideal than *absolute replacement*, in which sentient animals are not used at all (Russell & Burch 1959). *Intra-experimental reduction* is applied at the level of the experiment directly. *Supra-experimental reduction* is achieved through changes to the experimental or institutional setting. *Extra-experimental reduction* occurs via more distantly related developments (De Boo & Hendriksen 2005). Additional Rs occasionally proposed include *re-use* or *recycling* of animals to reduce total numbers, and *rehabilitation* to ensure the care and protection of animals after their laboratory use has ended (Anon. 1986, Pereira & Tettamanti 2005)

absolute replacement – see *3Rs*

ADME – see *-kinetics*

aetiology – study of causation, for example, of disease

altruistic behaviour – assistance of genetically unrelated others (although they are often members of the group to which the individual belongs) despite the absence of direct personal gain, and the possibility of incurring costs

analgesic – painkiller

antagonism – blockage or reversal of an effect, for example, of an *agonist* such as a pharmaceutical acting on a cellular receptor, by an *antagonist*

antigen – any *exogenous* substance evoking an immune response, either alone or after complexing with a larger molecule, that is capable of binding with a product (such as an antibody or specifically sensitised T lymphocyte) of that evoked immune response

apoptosis – cell death resulting from self-destruction; an important part of various normal tissue development processes. See also *neoplasia*

applicability domain of a toxicity test – spectrum of toxins detected by the test

arithmetic mean – see *mean*

Artiodactyla – taxonomic order including pigs, goats, sheep, and cattle

assay – a test, particularly to determine the biochemical or immunological activity of a sample. See also *bioassay*

bioassay – an animal test; commonly for toxicity. See also *assay*

biomarker of toxicity – biological molecule indicative of toxicity that can be detected, for instance, in blood or cell culture solutions; examples include induced detoxifying enzymes, cellular products, and chromosomal aberrations

biotransformation – see *metabolism*

blinding (of or pertaining to an experimental design) – prevents investigators or subjects from knowing the hypotheses or conditions being tested, thereby eliminating potential sources of bias from the results. For example, induction of pathology, experimental treatment, and outcome assessment should all be blinded

bolus – concentrated mass of a pharmaceutical or food

carcinogen – agent or influence causing cancer

CD28-SuperMAB – see *TGN1412*

cDNA microarray (gene chip) – contains hundreds or thousands of microscopic spots of cDNA transcripts of mRNA templates. Microarrays allow examination of the expression of large numbers of genes simultaneously. They can be used, for example, in *toxicogenomic* studies. See also *complementary DNA*

cell cultures – may be sourced from donor tissues directly (*primary cell cultures*) or from *cell lines*. *Immortalised cell lines* are capable of extended, and often indefinite, growth *in vitro*. See also *organ cultures; stem cell lines*

cell lines – see *cell cultures*

cephalopods – marine molluscs including octopuses, squid, and nautiloids

chi square (χ^2) test – a statistical calculation of the probability that two data sets are samples from the same underlying data population, and that any observed differences are simply due to random sampling variation. Large chi square values reflect increased probabilities that observed differences result from real differences in underlying data populations

cognition – higher mental processes such as the formation of associations, concepts, and insight, whose existence can only be inferred and not directly observed; in particular, processes involved in knowledge acquisition

comorbidity – concurrent pathology

complementary DNA (cDNA) segment – one from which non-coding *intron* sequences of the original DNA have been excised. See also *cDNA microarray*

confidence interval (CI) – range of values within which the true value of an estimated population parameter is calculated to lie with a specified probability. For example, the true value is 95 per cent certain to lie within a 95 per cent confidence interval

consistency approach in vaccine production – aims to achieve consistency among vaccine batches through compliance with good manufacturing practice and quality assurance principles, assisted by assessment at various stages using non-animal methods such as physicochemical methods and *in vitro* assays. This facilitates decreased animal testing of individual batches

conspecific – member of the same species

corticosterone – a stress hormone secreted by the adrenal cortex

cutaneous – pertaining to the skin. A *subcutaneous* injection inserts a substance into the region beneath the skin. See also *dermis*

-cyte – word element from the Greek word for 'cell'; for example, an *hepatocyte* is a liver cell

cytokine – regulatory protein which may be released by a cell population on contact with a specific *antigen*. Acts as an intercellular mediator, e.g. in the generation of an immune response

cytotoxin – antibody or other *endogenous* or *exogenous* compound exerting a toxic effect on certain cells

dam – female parent

decapods – order of crustaceans including prawns, shrimp, crabs, and lobsters

degrees of freedom (*df*) – number of independent pieces of information involved in calculating a parameter estimate. For the two-dimensional chi square arrays (with both rows and columns) analysed in Chapter 6, for example, the number of degrees of freedom associated with the resultant chi square values is calculated as (the number of rows – 1) × (the number of columns – 1)

demonstration experiment – one conducted to demonstrate established scientific concepts, such as physiological or biochemical principles, for teaching purposes

dermis – a layer within the skin. See also *cutaneous*

distress – see *stress*

-dynamics – word element; as in *pharmacodynamics* (PD) or *toxicodynamics*, the study of drug or toxin mechanisms of action and the resultant biochemical and physiological effects. See also *-kinetics*

embryonic stem cells (ESCs) – see *stem cell lines*

endogenous – of internal origin; for example, biologically active compounds such as hormones. Opposite of *exogenous*

epidemiology – study of the incidence, distribution, and control of diseases in large populations, and particularly of *aetiological* agents or risk factors

ethically sourced cadaver – one obtained from an animal that has been euthanased for medical reasons, or, less commonly, that has died naturally or in an accident; may be used in humane clinical or surgical training programmes

eukaryotic expression vector – DNA-containing molecule such as a plasmid or virus used as a vehicle to transfer foreign genetic material into a *eukaryotic* (nucleated) cell so that it can be expressed by the target cell. See also *recombinant shuttle vector*

euthanasia – a 'good death', that is, one clearly in the individual's best interests, ideally administered without pain or suffering. This normally occurs when the only alternative is severe and intractable suffering as a result of injury or disease. However, the term is often abused, being used to cover a wide range of deaths that are not in animals' best interests. Whether intentionally or otherwise, this abuse of the term can misrepresent the ethical legitimacy of such killing

evidence-based medicine (EBM) – bases clinical decisions on methodologically sound, prospective, randomised, blinded, controlled clinical trials. The gold standard for EBM is the large prospective epidemiological study, or *meta-analysis* of randomised and blinded, controlled clinical trials (Evidence-Based Medicine Working Group 1992). Similarly, *evidence-based toxicity assessment* relies on rigorous and impartial examination of evidence

evidence-based toxicity assessment – see *evidence-based medicine*

excitotoxicity – excessive stimulation of nerve cell receptors by excitatory neurotransmitters such as glutamate, resulting in cellular damage or death

exogenous – of external origin; for example, biologically active compounds such as pharmaceuticals or toxins. Opposite of *endogenous*

extra-experimental reduction – see *3Rs*

ex vivo – see *in vivo*

false belief understanding – awareness of the difference between reality and mental representations, and that the latter may be false. Considered a core criterion for advanced *Theories of Mind*

false negative (rate) – see *predictivity of a test*

false positive (rate) – see *predictivity of a test*

fidelity (of an animal model of humans) – accuracy with which an animal model reproduces human characteristics and responses. The *high fidelity fallacy* described by Russell and Burch (1959) is the incorrect assumption that, because placental mammals are similar to humans in some respects, they will always accurately reproduce key human responses, and hence will always provide the best possible model for medical research and testing

gastrotomy – surgical incision into the stomach

gavaging (usually orogastric gavaging) – insertion of a tube into the oesophagus for the forced administration of test compounds, for example, during toxicity tests

gene chip – see *cDNA microarray*

genetically altered organism – one in which the genome has been modified, whether artificially (*genetically modified*) or through conventional breeding or natural mutation

genetically modified (GM) organism – one in which the genome has been artificially altered; that is, not by conventional breeding or natural mutation. See also *genetically altered organism; knockout organism; transgenic organism*

genotoxin – compound capable of damaging DNA and thereby causing mutations or *neoplasia*

genotype – see *phenotype*

gross anatomy – macroscopic anatomy. See also *histology*

haptic technology – simulates tactile feedback that would be experienced, e.g. when manipulating real tissue. An important component of many virtual reality simulators, including those for teaching surgical or other clinical skills

hepatic – pertaining to the liver

hepatocyte – liver cell

High Production Volume Challenge Program (HPV) – EPA toxicity assessment programme targeting chemicals produced or imported into the US in quantities exceeding 1 million pounds annually

histology – microscopic anatomy of tissues. See also *gross anatomy*

homeostasis – see *stress*

humanisation of a biological product – protein-engineering of a product originating in a non-human species to ensure it expresses human-specific characteristics. See, for example, *TGN1412*

hypermedia – interactive information media in which graphics, audio, video, plain text, and hyperlinks intertwine in a structure that is generally non-linear. The broader term *multimedia* may be used to describe non-interactive linear presentations reliant on a variety of media, as well as hypermedia (Nelson 1965)

hypertension – abnormally high arterial blood pressure, or the resultant *systemic* disease syndrome

immortalised cell lines – see *cell cultures*

immunocompromisation – decreased competence of the immune system (*immunocompetence*), increasing disease predisposition. May result from factors such as stress, concurrent pathology, or *exogenous* steroids

inducible enzyme – an enzyme normally present in minute quantities in a cell whose concentration increases dramatically when a substrate compound is added, thereby facilitating adaptation to changing conditions

infarction – tissue death resulting from failure of blood supply. See also
ischaemia

in silico – see *in vivo*

Integrated Risk Information System (IRIS) chemicals database – EPA database containing known toxicity data, as well as human carcinogenicity and
other toxicity assessments, for the environmental contaminants of greatest
public health concern in the US

intra-experimental reduction – see *3Rs*

intraperitoneal – see *peritoneum*

invasive procedure – one interfering with bodily integrity, whether through
puncture or incision, or insertion of an instrument or foreign material, as
in surgical and some experimental procedures. *Markedly invasive* procedures
include those resulting in death (whether or not the subjects are conscious),
surgical procedures (other than very minor), major physiological challenges,
and the production of *genetically modified* animals

in vitro – see *in vivo*

in vivo – occurring in a living organism. Other processes occur *ex vivo*. These
may occur *in vitro* (as with cells or tissues grown in culture) or *in silico* (as in
computer simulations)

ischaemia – deficiency of blood supply to a body part caused by obstruction of
arterial inflow, for example, via blockage, rupture, or spasm. When the brain
is affected a *stroke* results. See also *infarction*

-kinetics – word element; as in *pharmacokinetics* (PK) or *toxicokinetics*; the study
of the movement of substances through the body and bodily compartments.
Its phases may include absorption, distribution, tissue localisation, biotransformation (*metabolism*), and excretion (ADME). See also *-dynamics*

knockout organism – one in which one or more genes have been deleted or
inactivated. See also *genetically modified organism*

LD50 (Lethal Dose 50) test – administration (frequently by orogastric *gavaging*)
of increasing doses of a test substance to determine the dose lethal to 50 per
cent of a laboratory animal group

lesion – any abnormal, localised change in the structure of an organ or body
part as a result of injury or disease

longitudinal study – one in which observations are repeated over a prolonged
period. In some studies animals may act as their own controls, reducing animal numbers required

marginal human person – see *personhood*

markedly invasive procedure – see *invasive procedure*

maximum tolerated dose (MTD) – maximum possible dose, for example, of a
toxin, above which acute toxicity-related effects become intolerable

mean – one of several measures of central tendency, or averages, used in descrip-
tive statistics. For an *arithmetic mean*, each data point contributes equally to
the final result; for a *weighted mean*, some contribute more than others

meta-analysis – aggregation and statistical analysis of suitable data from multiple experiments

metabolism (biotransformation) – sum of the physical and chemical processes
by which living tissue is built up and maintained (*anabolism*), and by which
large molecules are broken down into smaller molecules to make energy available (*catabolism*). The biological activity and toxicity of *metabolites* produced

by metabolic pathways and enzyme systems may differ notably from those of their precursors

metabolite – see *metabolism*

metastasis – transfer of a disease such as *neoplasia* from one organ or body part to another not directly connected to it

microdose (phase 0) study – A first-in-human study. Phase 0 studies use doses estimated from preclinical studies to be well below thresholds believed necessary for pharmacologic or toxicological activity. They are conducted to provide information on human bioavailability and pharmacokinetics

mitogenesis – induction of cellular division resulting in mitosis. This may lead to physiological or pathological cellular proliferation. See also *neoplasia*

multimedia – see *hypermedia*

multipotent adult progenitor cells – see *stem cell lines*

murine – belonging or pertaining to the Muridae family of rodents, which includes mice and rats

negative predictivity – see *predictivity of a test*

neoplasia – cellular proliferation, often uncontrolled and associated with a failure of *apoptosis*; frequently refers to the malignant proliferation of cancer cells, which in some cancers can result in *metastasis*. See also *oncogene*

neuromuscular blocking agent – pharmaceutical inducing muscular paralysis; sometimes used to induce immobility during clinical or surgical procedures, but does not normally have anaesthetic or analgesic activity, so the subject may remain normally aware and capable of sensation. Reliance solely on a neuromuscular blocker for most invasive procedures is therefore unethical and inhumane, and may be prohibited

New World monkeys – include the capuchin, douroucouli, howler monkey, marmoset, saki, spider monkey, squirrel monkey, titi, uakari, and woolly monkey

nociceptor – receptor sensitive to painful or injurious stimuli, such as mechanical, chemical, electrical, and thermal insults

non-human person – see *personhood*

non-recovery procedure – see *terminal procedure*

no observed adverse effect level (NOAEL) – maximum concentration or dose of a compound at which no adverse changes are detectable

odds ratio (OR) – measure of effect size which describes the strength of association or non-independence between two binary data values

Old World monkeys – include the baboon, colobus monkey, guenon, langur, macaque, mandrill, mangabey, patas, proboscis, and talapoin

oncogene – gene carried by a tumour virus or cancer cell which is solely or partly responsible for *tumorigenesis* (tumour formation). See also *neoplasia*

optimism bias – favourable consideration of data that appear to confirm a given hypothesis, in contrast to data that appear to refute it. For example, animal data apparently confirming the hypothesised human clinical benefit of a test pharmaceutical compound or other medical intervention are often given greater weight than animal data of similar scientific merit which refute such a hypothesis. This may lead to *publication bias*

organ cultures – three-dimensional *cell cultures* that retain some or all of the *histological* features of the *in vivo* equivalent

ovariohysterectomy – surgical removal of the uterus and ovaries

parenteral drug – one delivered by a *non-enteric* (non-intestinal, i.e. not oral) route of administration

pathogenesis – processes involved in the development of disease

pathophysiology – physiology of disordered functioning

peptide – an amide derived from two or more amino acids by combination of the amino group of one acid with the carboxyl group of another; usually obtained by partial hydrolysis of proteins

peritoneum – serous membrane that lines the abdominal cavity and encloses its *viscera* (internal organs). The *peritoneal cavity* is the abdominal cavity, and is the target of an *intraperitoneal* injection

Perissodactyla – taxonomic order including horses, donkeys, and crossbreeds

personhood – although such anthropocentrism has recently been challenged, classical concepts of personhood rely on the possession of human-like psychological characteristics, such as consciousness and self-consciousness; the capacity to experience a wide range of emotional states; and the possession of key *cognitive* abilities, including those giving rise to culture, language, and a *Theory of Mind*. *Marginal human persons* include the very young, old, injured, or ill, who lack the full range of psychological and social characteristics and abilities exhibited by healthy human adults. They are nevertheless valued as partially conscious, partially self-conscious, or partially autonomous beings with unique personalities, and accordingly are normatively and legislatively considered to have human rights. *Non-human persons* may possess a similar array of morally relevant characteristics

pharmacodynamics – see *-dynamics*

pharmacokinetics – see *-kinetics*

phase 0 study – see *microdose study*

phenotype – outward appearance of an animal, encompassing all its anatomical, physiological, and behavioural characteristics; determined by both genetic and environmental factors. Phenotype contrasts with *genotype*, which is limited to inherited factors

phylogeny – evolutionary history of a group of organisms, or race

pleural cavity – thoracic cavity

pluripotent embryonic stem cells (ESCs) – see *stem cell lines*

poikilotherm – organism whose internal temperature varies considerably. In laboratory animal use statistics, this term often refers to fish, amphibians, and reptiles

positive predictivity – see *predictivity of a test*

power of an experiment – see *p-value*

practical validation – see *validation*

predictivity of a test – *positive predictivity* is the proportion of positive test outcomes that are truly positive for the characteristic being tested for (*true positives*); the *false positive rate* refers to the proportion that are not (*false positives*). The *negative predictivity* and *false negative rates* are similarly determined. Predictivities vary with the prevalence of the characteristic in the population being tested, whereas test *sensitivity* and *specificity* are fixed and determined only by the test

primary cell cultures – see *cell cultures*

prolactin – hormone stimulating milk production and corpus luteum maintenance in mammals, crop milk production in certain birds, and other effects in lower vertebrates

prosection – expertly performed dissection to demonstrate anatomy

prosimian – the lemur, loris, potto, bush baby, and aye-aye

publication bias – an increased tendency to publish studies in which a treatment effect is apparent, or a decreased tendency to publish such studies, which may result, for example, from commercial pressures, particularly in the case of patented drugs under development

p-value or α – probability of assuming a difference where none exists (a *Type I error*). An apparent difference is normally considered statistically significant when $p < 0.05$. The opposite term is β, the probability of assuming no difference where one does exist (a *Type II error*). The *power* of an experiment is calculated as $1 - \beta$. Experimental parameters such as sample size are most commonly chosen to ensure the power is at least 0.8

pyrogen – fever-inducing compound

quantitative structure–activity relationship (QSAR) – mathematical description of the relationship between the physicochemical properties of a molecule and its biological activities. See also *structure–activity relationship*

read-across – prediction of unknown characteristics of compounds such as physicochemical properties, environmental fates, and toxicological effects from those of other compounds which share commonalities of physicochemical structure or metabolic pathways, and hence may be grouped in the same chemical category. May be used in toxicity assessment

recombinant shuttle vector – vector designed to express genetic material in cells of two different species. Such vectors must carry the genes necessary for expression not provided by each host cell, and are created by DNA recombination (*gene splicing*). See also *eukaryotic expression vector*

recycling of laboratory animals – see *3Rs*

reduction of laboratory animal use – see *3Rs*

refinement of laboratory animal use – see *3Rs*

Registration, Evaluation and Authorisation of Chemicals (REACH) – EC programme that aims to assess the toxicity of some 30,000 chemicals produced or imported into the EU in quantities in excess of 1 metric ton annually

rehabilitation of laboratory animals – see *3Rs*

relative replacement – see *3Rs*

renal – pertaining to the kidneys

replacement of laboratory animal use – see *3Rs*

retrospective analysis of experimental protocols – allows assessment of the degree to which experimental objectives were successfully met, and of the extent to which animal suffering matched that predicted in licensing applications. It may inform future research design and licensing decisions

re-use of laboratory animals – see *3Rs*

reverse toxicology approach – positions hazard characterisation and assessment after exposure assessment during the toxicological risk assessment of a compound. This reverses the order in which such assessment traditionally occurs, and allows modulation of toxicity assessment requirements – including for animal testing – on the basis of human exposure risks

sensitivity of a test – likelihood of correctly detecting compounds possessing the property of interest (*true positives*). Failure to detect these results in *false negatives*. The ideal test has maximal sensitivity and specificity, but an increase in one often results in a decrease in the other. See also *predictivity*; *specificity*

specificity of a test – likelihood of correctly identifying compounds lacking the property of interest (*true negatives*). Failure to correctly identify these results in *false positives*. See also *predictivity; sensitivity*

stem cell lines – may be established from mammalian blastocysts (*pluripotent embryonic stem cells*, which can differentiate into all cell types other than additional embryonic tissue) or from *multipotent adult progenitor cells* (which have the potential to differentiate into cells from a limited number of lineages). Recent evidence suggests that the differentiation potential of adult stem cells may not be as limited as previously thought

stereotypical behaviours – repetitive, apparently purposeless behaviours, believed to indicate psychological distress which is both profound and chronic

stress – 'the effect produced by external (i.e., physical or environmental) events or internal (i.e., physiologic or psychologic) factors…which induce an alteration in an animal's biologic equilibrium' (i.e. state of *homeostasis*). *Stressors* are adverse stimuli resulting in stress, and *distress* is 'an aversive state in which an animal is unable to adapt completely to stressors and the resulting stress and shows maladaptive behaviors' (ILAR 1992)

stressor – see *stress*

stroke – see *ischaemia*

structure–activity relationship (SAR) – relationship between physicochemical structure and biological activity. See also *quantitative structure–activity relationship*

subcutaneous – see *cutaneous*

supra-experimental reduction – see *3Rs*

systematic review (of the human clinical or toxicological utility of animal experiments) – one in which the experiments reviewed are selected randomly or via other impartial and methodical means to minimise bias. To ensure statistical significance of any results, large numbers of experiments are normally reviewed

systemic – relating to or affecting the entire body, rather than a localised region

teratogen – an agent or influence causing physical defects in the developing embryo

terminal procedure – one resulting in the death of the subject

TGN1412 (CD28-SuperMAB) – fully *humanised* monoclonal antibody that was being developed for the treatment of inflammatory conditions. During a 2006 phase I clinical trial in the UK, TGN1412 caused severe adverse reactions in all six volunteers given the drug, with one volunteer suffering permanent damage

Theory of Mind (ToM) – those possessing a ToM are considered able to ascribe mental or psychological states to themselves and others, including perceptual states such as seeing, as well as beliefs and desires (Premack & Woodruff 1978, Bischof-Köhler 2000). ToM is a core consciousness-based capacity of human beings. Some evidence also points to the existence of a ToM in certain other species (Benz-Schwarzburg & Knight 2011)

Three Rs – see *3Rs*

tissue cultures – *in vitro* cultivations of cells, organs, embryos, or other tissues. Cultures may be *static* or *perfused* with culture media via pumps. See also *cell cultures; organ cultures*

toxicodynamics – see *-dynamics*

toxicogenomics – application of *genomics* (the study of genomes) to toxicology. For example, *cDNA microarrays* can be used to link changes in genetic expression with exposure to certain classes of toxic compounds

toxicokinetics – see *-kinetics*

transgenic organism – one in which one or more foreign genes have been inserted into the genome. See also *genetically modified organism*

true negative – see *predictivity of a test*

true positive – see *predictivity of a test*

utilitarianism – philosophical doctrine according to which the morality of a choice or action is determined by the resultant change in utility (such as pleasure, preference satisfaction, or useful knowledge) for all sentient beings affected. A *utilitarian cost:benefit analysis* fundamentally underpins most modern regulations governing animal experimentation. The likely benefits – normally for human beings – must exceed the probable harms to the animal subjects

validation, as in scientific validation of an alternative assay – traditionally, the demonstration in multiple independent laboratories that the *assay* is relevant and reliable for its specified purpose (*practical validation*; Balls et al. 1995). However, it is not always scientifically necessary, or even logistically possible, to conduct multi-centre practical studies. Hence *weight-of-evidence validation*, or *validation by retrospective analysis* (ICCVAM 1997, OECD 2003), may be conducted, which aims to review all available, relevant evidence in a structured, systematic, independent, and transparent manner – provided sufficient data are available (Balls & Combes 2006)

venipuncture – blood sampling from a vein

weighted mean – see *mean*

weight-of-evidence validation or assessment – see *validation*

xenobiotic – foreign to a living organism; for example, drugs or toxins

References

Aardema M & MacGregor JT (2002). Toxicology and genetic toxicology in the new era of 'toxicogenomics': impact of 'omics' technologies. *Mutat Res* 499: 13–25.

Aarstad HJ & Seljelid R (1992). Effects of stress on the growth of a fibrosarcoma in nu/nu and conventional mice. *Scand J Immunol* 35: 209–15.

Abutarbush SM *et al.* (2006). Evaluation of traditional instruction versus a self-learning computer module in teaching veterinary students how to pass a nasogastric tube in the horse. *J Vet Med Educ* 33(3): 447–54.

Akyüz N & Wiesmüller L (2003). Proof of principle: detection of genotoxicity by a fluorescence-based recombination test in mammalian cells. *ALTEX* 20(2): 77–84.

Allen SW & Chambers JN (1997). Computer-assisted instruction of fundamental surgical motor skills. *J Vet Med Educ* 24(1): 2–5.

Amacher DE & Zelljadt I (1983). The morphological transformation of Syrian hamster embryo cells by chemicals reportedly nonmutagenic to *Salmonella typhimurium. Carcinogenesis* 4: 291–5.

Andersen ME (1991). Physiological modelling of organic compounds. *Ann Occup Hyg* 35: 309–21.

Anon. (1986). On the re-use of laboratory animals. *Altern Lab Anim* 13(3): 153–5.

Anon. (2000). *Characterisation of Data Uncertainty and Variability in IRIS Assessments: Pre-pilot vs. Pilot/Post-pilot.* Springfield, VA: Versar Inc.

Anon. (2001). *2001 Veterinary Student Survey Report.* Palmerston North, New Zealand: Massey University. [unpublished]. www.humanelearning.info/resources/surveys.htm, accessed 25 Jun. 2010.

Anon. (2004a). Re-homing primates: the consequences of zero-tolerance to primate research. *Altern Lab Anim* 32(3): 153–4.

Anon. (2004b). Nearly every veterinary school cited by USDA for non-compliance with federal law. *Altern Vet Med Educ* 25: 2–3.

Anon. (2006). [Name withheld by request as this University of Sydney veterinary student had not yet graduated]. Personal communication to Andrew Knight re: faculty opposition to humane teaching methods and limited compliance with safety guidelines relating to chemically preserved anatomy specimens at the University of Sydney Faculty of Veterinary Science.

Anon. (2007). End great ape experiments, says European Parliament. *New Sci* 2621: 4.

Arlt S & Heuwieser W (2005). Evidenz-Basierte Veterinarmedizin. [German]. [Evidence based veterinary medicine]. *Deutsche Tierärztliche Wochenschrift* 112: 146–8.

Armstrong SJ & Botzler RG (eds) (2003). *The Animal Ethics Reader.* London: Routledge.

Aronowski J, Strong R & Grotta JC (1996). Treatment of experimental focal ischemia in rats with lubeluzole. *Neuropharmacology* 35: 689–93.

Ashby J (1996). Alternatives to the two-species bioassay for the identification of potential human carcinogens. *Hum Exp Toxicol* 15: 183–202.

Ashby J & Purchase IF (1993). Will all chemicals be carcinogenic to rodents when adequately evaluated? *Carcinogenesis* 8: 489–95.

Ashby J & Tennant RW (1991). Definitive relationships among chemical structure, carcinogenicity and mutagenicity for 301 chemicals tested by the U.S. NTP. *Mutat Res* 257: 229–306. Erratum in *Mutat Res* 1994; 317: 175.

AVMA (2005). (American Veterinary Medical Association). About the American Veterinary Medical Association. www.avma.org/membshp/about.asp, accessed 21 Jul. 2005.

AVMA (2011). (American Veterinary Medical Association). Animal welfare policy statements. www.avma.org/issues/animal_welfare/policies.asp, accessed 17 Mar. 2011.

AWIC (2011). (Animal Welfare Information Center). Animal Welfare Act. http://awic.nal.usda.gov/, accessed 21 Apr. 2011.

Bailey J (2005). Non-human primates in medical research and drug development: a critical review. *Biog Amines* 19(4–6): 235–55.

Bailey J (2008). An assessment of the role of chimpanzees in AIDS vaccine research. *Altern Lab Anim* 36: 381–428.

Bailey J (2009). An examination of chimpanzee use in human cancer research. *Altern Lab Anim* 37: 399–416.

Bailey J (2010). An assessment of the use of chimpanzees in hepatitis C research past, present and future: 1. validity of the chimpanzee model. *Altern Lab Anim* 38(5): 387–418.

Bailey J, Knight A & Balcombe J. (2005). The future of teratology research is *in vitro*. *Biog Amines* 19, 97–145.

Baillie S *et al.* (2003). Preliminary development and evaluation of a bovine rectal palpation simulator for training veterinary students. *Cattle Pract* 11(2): 101–6.

Baillie S *et al.* (2005a). Validation of a bovine rectal palpation simulator for training veterinary students. *Stud Health Technol Inform* 111: 33–6.

Baillie S *et al.* (2005b). Integrating a bovine rectal palpation simulator into an undergraduate veterinary curriculum. *J Vet Med Educ* 32(1): 79–85.

Balcombe J (1998). Letter to the Editor. *Am Biol Teach* 60(8): 555–6.

Balcombe J (2000a). A global overview of law and policy concerning animal use in education. In M Balls, A-M Zeller & M Halder (eds). *Progress in the Reduction, Refinement and Replacement of Animal Experimentation.* New York: Elsevier. 1343–50.

Balcombe J (2000b). *The Use of Animals in Higher Education: Problems, Alternatives, and Recommendations.* Washington DC: Humane Society Press.

Balcombe J (2006). Laboratory environments and rodents' behavioural needs: a review. *Lab Anim* 40: 217–35.

Balcombe J, Barnard N & Sandusky C (2004). Laboratory routines cause animal stress. *Contemp Top Lab Anim Sci* 43: 42–51.

Baldwin A & Bekoff M (2007). Too stressed to work. *New Sci* 194(2606): 24.

Balls M (1994). Replacement of animal procedures: alternatives in research, education and testing. *Altern Lab Anim* 28: 193–211.

Balls M (1995). Chimpanzee medical experiments: moral, legal, and scientific concerns. In *Poor Model Man: Experimenting on Chimpanzees: Proceedings of* the First PACE (People Against Chimpanzee Experiments) Conference on the Use of *Chimpanzees in Biomedical Research*. *Altern Lab Anim* 23: 607–14.

Balls M (2004). Are animal tests inherently valid? *Altern Lab Anim* 32(Suppl. 1B): 755–8.

Balls M & Combes R (2005). The need for a formal invalidation process for animal and non-animal tests. *Altern Lab Anim* 33: 299–308.

Balls M & Combes R (2006). Validation via weight-of-evidence approaches. *ALTEX* 23(Sp. Issue): 332–5.

Balls M, Festing MFW & Vaughan S (eds) (2004). Reducing the use of experimental animals where no replacement is yet available. *Altern Lab Anim* 32(Suppl. 2): 1–104.

Balls M *et al.* (1990). Report and recommendations of the CAAT/ERGATT workshop on the validation of toxicity test procedures. *Altern Lab Anim* 18: 313–37.

Balls M *et al.* (1995). Practical aspects of the validation of toxicity test procedures: the report and recommendations of ECVAM Workshop 5. *Altern Lab Anim* 23: 129–47.

Balluch M (2006). Animals have a right to life. *ALTEX* 23(4): 281–6.

Balluch M (2007). Trial for personhood of a chimp. *ALTEX* 24(3): 186.

Balmain A & Harris CC (2000). Carcinogenesis in mouse and human cells: parallels and paradoxes. *Carcinogenesis* 21: 371–7.

Bauer MS (1993). A survey of the use of live animals, cadavers, inanimate models, and computers in teaching veterinary surgery. *J Am Vet Med Assoc* 203(7): 1047–51.

Bauer MS *et al.* (1992). Evaluation of the effectiveness of a cadaver laboratory during a fourth-year veterinary surgery rotation. *J Vet Med Educ* 19(2): 77–84.

Bebarta V, Luyten D & Heard K (2003). Emergency medicine animal research: does use of randomisation and blinding affect the results? *Acad Emerg Med* 10: 684–7.

Benigni R, Passerini L & Rodomonte A. (2003). Structure-activity relationships for the mutagenicity and carcinogenicity of simple and alpha-beta unsaturated aldehydes. *Environ Mol Mutagen* 42: 136–43.

Benz-Schwarzburg J & Knight A (2011). Cognitive relatives yet moral strangers? *J Anim Ethics* 1(1): 9–36.

Beran MJ (2004). Long-term retention of the differential values of Arabic numerals by chimpanzees (*Pan troglodytes*). *Anim Cogn* 7(2): 86–92.

Besaratinia A, Kleinjans JC & Van Schooten FJ (2002). Biomonitoring of tobacco smoke carcinogenicity by dosimetry of DNA adducts and genotyping and phenotyping of biotransformational enzymes: a review on polycyclic aromatic hydrocarbons. *Biomarkers 7*, 209–29.

Bettauer RH (2010). Chimpanzees in hepatitis C virus research: 1998–2007. *J Med Primatol* 39: 9–23.

Bhogal N & Combes R (2006). TGN1412: time to change the paradigm for the testing of new pharmaceuticals. *Altern Lab Anim* 34: 225–39.

Bischof-Köhler D (2000). *Kinder auf Zeitreise: Theory of Mind, Zeitverständnis und Handlungsorganisation*. [German]. [*Children on a Time Trip. Theory of Mind, the Concept of Time and the Organization of Actions*]. Bern, Switzerland: Huber.

Blaauboer BJ, Hermens J & Van Eijkeren J (2006). Estimating acute toxicity based on *in vitro* cytotoxicity: role of biokinetic modelling. *ALTEX* 23(Sp. Issue): 250–3.

Blum MD, Graham DJ & McCloskey CA (1994). Temafloxacin syndrome: review of 95 cases. *Clin Infect Dis* 18: 946–50.

Bourgeois SR, Vazquez M & Brasky K (2007). Combination therapy reduces self-injurious behavior in a chimpanzee (*Pan Troglodytes Troglodytes*): a case report. *J Appl Anim Welf Sci* 10(2): 123–40.

Bradshaw GA *et al.* (2008). Building an inner sanctuary: complex PTSD in chimpanzees. *J Trauma Dissociation* 9(1): 9–34.

Bremer S *et al.* (2007). The development of new concepts for assessing reproductive toxicity applicable to large scale toxicological programmes. *Curr Pharm Des* 13(29): 3047–58.

Brenner GJ *et al.* (1990). Increased pulmonary metastases and natural killer cell activity in mice following handling. *Life Sci* 47: 1813–19.

Brent L (2004). Solutions for research chimpanzees. *Lab Anim* 33(1): 37–43.

Britten RJ (2002). Divergence between samples of chimpanzee and human DNA sequences is 5%, counting indels. *Proc Natl Acad Sci* 99(21): 13633–5.

Broadhead C & Bottrill K (1997). Strategies for replacing animals in biomedical research. *Mol Med Today* 3(11): 483–7.

Brom FW (2002). Science and society: different bioethical approaches towards animal experimentation. *ALTEX* 19: 78–82.

Brown CM *et al.* (1995a). Neuroprotective properties of lifarizine compared with those of other agents in a mouse model of focal cerebral ischaemia. *Br J Pharmacol* 115: 1425–32.

Brown NA (1987). Teratogenicity testing *in vitro*: status of validation studies. *Arch Toxicol Suppl* 11: 105–14.

Brown NA *et al.* (1995b). Screening chemicals for reproductive toxicity: the current alternatives. *Altern Lab Anim* 23: 868–82.

Brownlee S (2000, 15 May). Surplus chimps stranded in research controversy. *Washington Post*: A9.

Brüne M *et al.* (2006). Psychopathology in great apes: concepts, treatment options and possible homologies to human psychiatric disorders. *Neurosci Biobehav Rev* 30: 1246–59.

Brusick DJ (1977). *In vitro* mutagenesis assays as predictors of chemical carcinogenesis in mammals. *Clin Toxicol* 10: 79–109.

Buchanan MF *et al.* (2005). Using 3D animations to teach intracellular signal transduction mechanisms: taking the arrows out of cells. *J Vet Med Educ* 32(1): 72–8.

Buchanan-Smith HM *et al.* (2005). Harmonising the definition of refinement. *Anim Welf* 14: 379–84.

Buesen R *et al.* (2004). Trends in improving the embryonic stem cell test (EST): an overview. *ALTEX* 21(1): 15–22.

Burkart JM *et al.* (2007). Other-regarding preferences in a non-human primate: common marmosets provision food altruistically. *Proc Natl Acad Sci USA* 104(50): 19762–6.

Busche F (2008). Animal rights, animal wrongs. *Vet Times* 38(16): 32–4.

Butler AB (2008). Evolution of the thalamus: a morphological and functional review. *Thalamus Relat Syst* 4: 35–58.

Cahill PA *et al.* (2004). The GreenScreen genotoxicity assay: a screening validation programme. *Mutagenesis* 19: 105–19.

Call J (2003). Beyond learning fixed rules and social cues: abstraction in the social arena. *Philos Trans R Soc Lond B Biol Sci* 358(1435): 1189–96.

Call J & Tomasello M (2008). Does the chimpanzee have a theory of mind? 30 years later. *Trends Cogn Sci* 12(5): 187–92.

Callaham M, Wears RL & Weber E (2002). Journal prestige, publication bias, and other characteristics associated with citation of published studies in peer-reviewed journals. *J Amer Med* 287: 2847–50.

Canfield P & Taylor R (2005). Teaching and learning at the Faculty of Veterinary Science, University of Sydney. *J Vet Med Educ* 32(3): 349–58.

Carpenter LG *et al.* (1991). A comparison of surgical training with live anesthetized dogs and cadavers. *Vet Surg* 20(6): 373–8.

CAST (1997) (Chinese Acute Stroke Trial Collaborative Group). Randomised placebo-controlled trial of early aspirin use in 20,000 patients with acute ischaemic stroke. *Lancet* 349: 1641–9.

Cavalieri P & Singer P (eds) (1993). *The Great Ape Project*. London: Fourth Estate.

Caversaccio M, Eichenberger A & Hausler R (2003). Virtual simulator as a training tool for endonasal surgery. *Am J Rhinol* 17(5): 283–90.

CCAC (2009). (Canadian Council on Animal Care). *2008 CCAC Survey of Animal Use*. Ottawa: CCAC. http://ccac.ca/Documents/Publications/Statistics/Survey_2008.pdf, accessed 6 Feb. 2011.

CDER (2006) (Center for Drug Evaluation and Research, US Department of Health and Human Services, Food and Drug Administration). *Guidance for Industry, Investigators, and Reviewers: Exploratory IND Studies*. Rockville, MD: CDER.

Charles GD (2004). *In vitro* models in endocrine disruptor screening. *ILAR J* 45(4): 494–501.

Clarke KA (1987). The use of microcomputer simulations in undergraduate neurophysiology experiments. *Altern Lab Anim* 14: 134–40.

Clevenger J & Kass PH (2003). Determinants of adoption and euthanasia of shelter dogs spayed or neutered in the University of California veterinary student surgery program compared to other shelter dogs. *J Vet Med Educ* 30(4): 372–8.

Coghlan A (2006, 14 Aug.). Mystery over drug trial debacle deepens. *New Sci.* www.newscientist.com/article.ns?id=dn9734, accessed 13 Jun. 2010.

Cohen J (2007a). The endangered lab chimp. *Science* 315: 450–2.

Cohen J (2007b). NIH to end chimp breeding for research. *Science* 316: 1265.

Cohen PS & Block M (1991). Replacement of laboratory animals in an introductory psychology laboratory. *Humane Innovations and Alternatives* 5: 221–5.

Cohen SM (1995). Human relevance of animal carcinogenicity studies. *Regul Toxicol Pharmacol* 21: 75–80.

Cohen SM & Lawson TA (1995). Rodent bladder tumors do not always predict for humans. *Cancer Lett* 93: 9–16.

Comber MH *et al.* (2003). Quantitative structure-activity relationships for predicting potential ecological hazard of organic chemicals for use in regulatory risk assessments. *Environ Toxicol Chem* 22: 1822–8.

Combes RD (2007). Developing, validating and using test batteries and tiered (hierarchical) testing schemes. *Altern Lab Anim* 35: 375–8.

Combes RD *et al.* (1999). Cell transformation assays as predictors of human carcinogenicity: the report and recommendations of ECVAM Workshop 39. *Altern Lab Anim* 27: 745–67.

Combes RD *et al.* (2003). Early microdose drug studies in human volunteers can minimise animal testing – proceedings of a workshop organised by Volunteers in Research and Testing. *Eur J Pharm Sci* 19: 1–11.

Combes RD *et al.* (2004). The Third FRAME Toxicity Committee: working toward greater implementation of alternatives in toxicity testing. *Altern Lab Anim* 32(Suppl. 1B): 635–42.

Combes RD *et al.* (2007). Proposed integrated decision-tree testing strategies for mutagenicity and carcinogenicity in relation to the EU REACH legislation. *Altern Lab Anim* 35(2): 267–87.

Cone JE & Rosenberg J (1990). Medical surveillance and biomonitoring for occupational cancer endpoints. *Occup Med* 5: 563–81.

Conlee KM & Boysen ST (2005). Chimpanzees in research: past, present, and future. In DJ Salem and AN Rowan (eds). *The State of the Animals III: 2005.* Gaithersburg, MD: Humane Society Press. 119–37.

Conlee KM, Hoffeld EH & Stephens ML (2004). A demographic analysis of primate research in the United States. *Altern Lab Anim* 32(Suppl 1A): 315–22.

Conolly RB *et al.* (1988). Pharmacokinetics, biochemical mechanism and mutation accumulation: a comprehensive model of chemical carcinogenesis. *Toxicol Lett* 43: 189–200.

Corpet DE & Pierre F (2005). How good are rodent models of carcinogenesis in predicting efficacy in humans? A systematic review and meta-analysis of colon chemoprevention in rats, mice and men. *Eur J Cancer* 41: 1911–22.

Corry DB & Kheradmand F (2005). The future of asthma therapy: integrating clinical and experimental studies. *Immunol Res* 33: 35–51.

Cosson P (2007). A non-mammalian system to study bacterial infections. *ALTEX* 24(Sp. Issue): 78–9.

Council of Europe (1986). *European Convention for the Protection of Vertebrate Animals Used for Experimental and Other Scientific Purposes* (ETS 123). Strasbourg, France: Council of Europe.

Council of Europe (ed.) (2006). *Ethical Eye: Animal Welfare.* Belgium: Council of Europe Publishing.

Crespi C (1995). Xenobiotic-metabolizing human cells as tools for pharmacological and toxicological research. *Adv Drug Res* 26: 180–235.

Cronin MT *et al.* (2003). Quantitative structure-activity relationships for human health effects: commonalities with other endpoints. *Environ Toxicol Chem* 22: 1829–43.

Cross TR & Cross VE (2004). Scalpel or mouse, a statistical comparison of real and virtual frog dissections. *Am Biol Teach* 66(6): 408–11.

CSAC (2005) (Chimpanzee Sequencing and Analysis Consortium). Initial sequence of the chimpanzee genome and comparison with the human genome. *Nature* 437(7055): 69–87.

Cudilo E *et al.* (2007). Knockout mice: is it just genetics? Effect of enriched housing on fibulin-4+/- mice. *PLoS ONE* 2(2): e229.

Curry SH (2003). Why have so many drugs with stellar results in laboratory stroke models failed in clinical trials? A theory based on allometric relationships. *Ann N Y Acad Sci* 993: 69–74.

Dagg AI (1999). Responsible animal-based research: three flags to consider. *J Appl Anim Welf Sci* 2(4): 337–46.

Dagg AI (2000). Animal experimentation in cancer research: a citation analysis. *J Appl Anim Welf Sci* 3(3): 239–51.

Dagg AI & Seidle TK (2004). Levels of citation of nonhuman animal studies conducted at a Canadian research hospital. *J Appl Anim Welf Sci* 7(3): 205–13.

Dahl SL & Ward JR (1982). Pharmacology, clinical efficacy, and adverse effects of the nonsteroidal anti-inflammatory agent benoxaprofen. *Pharmacotherapy* 2: 354–66.

Dale VH, Johnston PE & Sullivan M (2003). Learning and teaching innovations in the veterinary undergraduate curriculum at Glasgow. *J Vet Med Educ* 30(3): 221–5.

De Boo J & Hendriksen C (2005). Reduction strategies in animal research: a review of scientific approaches at the intra-experimental, supra-experimental and extra-experimental levels. *Altern Lab Anim* 33: 369–77.

De Boo J & Knight A (2005). 'Concepts in animal welfare': a syllabus in animal welfare science and ethics for veterinary schools. *J Vet Med Educ* 32(4): 451–3.

De Boo J & Knight A (2006). Educating the veterinary professional about animal welfare. *ALTEX* 23(Sp. Issue: Proceedings: 5th World Congress 2005): 71–4.

De Boo J & Knight A (2008). Increasing the implementation of alternatives to laboratory animal use. *AATEX* 13(3): 109–17.

De Boo J *et al.* (2005). The interplay between replacement, reduction and refinement: considerations where the Three Rs interact. *Anim Welf* 14: 327–32.

De Veer MW *et al.* (2003). An 8-year longitudinal study of mirror self-recognition in chimpanzees (Pan troglodytes). *Neuropsychologia* 41(2): 229–34.

Dearfield KL *et al.* (1991). Considerations in the U.S. Environmental Protection Agency's testing approach for mutagenicity. *Mutat Res* 258: 259–83.

DeGrazia D (1996). *Taking Animals Seriously: Mental Life and Moral Status.* Cambridge: Cambridge University Press.

Department of Health (2000) (Committee on Mutagenicity of Chemicals in Food, Consumer Products and the Environment, Dept of Health, UK). *Guidance on a Strategy for Testing of Chemicals for Mutagenicity.* London: Dept of Health.

Dewhurst DG & Jenkinson L (1995). The impact of computer-based alternatives on the use of animals in undergraduate teaching. *Altern Lab Anim* 23: 521–30.

Dewhurst DG & Meehan AS (1993). Evaluation of the use of computer simulations of experiments in teaching undergraduate students. *Br J Pharm Proc Suppl* 108: 238.

Dewhurst DG, Brown GJ & Meehan A (1988). Microcomputer simulations of laboratory experiments in physiology. *Altern Lab Anim* 15(4), 280–90.

Dewhurst DG *et al.* (1994). Comparison of a computer simulation program and a traditional laboratory practical class for teaching the principles of intestinal absorption. *Am J Physiol* 267(6 Pt 3): 95–104.

Dhein CR & Memon M (2003). Online continuing education at the College of Veterinary Medicine, Washington State University. *J Vet Med Educ* (Sp. Issue: Continuing Veterinary Education) 30(1): 41–6.

Di Carlo FJ (1984). Carcinogenesis bioassay data: correlation by species and sex. *Drug Metab Rev* 15: 409–13.

DiMasi JA, Hansen RW & Grabowski HG (2003). The price of innovation: new estimates of drug development costs. *J Health Econ* 22(2): 151–85.

Docterman KE & Smith SM (2002). Of meis and men: lessons from a microarray study of teratogen action. *Teratology* 66: 217–23.

Doherty ML & Boyd RJ (2006). Undergraduate veterinary education at University College Dublin: a time of change. *J Vet Med Educ* 33(2): 214–19.

Downie R & Meadows J (1995). Experience with a dissection opt-out scheme in university level biology. *J Biol Educ* 29(3): 187–94.

Dybing E & Huitfeldt HS (1992). Species differences in carcinogen metabolism and interspecies extrapolation. In H Vainio *et al.* (eds) *Mechanisms of Carcinogenesis in Risk Identification.* Lyon, France: International Agency for Research on Cancer. 501–22.

Dyson DH (2003). Non-linear, visual-rich supplemental material designed for an introductory course in veterinary anaesthesia. *J Vet Med Educ* 30(4): 360–3.

EC (1999) (European Commission). *Second Report from the Commission to the Council and the European Parliament on the Statistics on the Number of Animals used for Experimental and other Scientific Purposes in the Member States of the European Union.* Brussels: EC. http://ec.europa.eu/environment/chemicals/lab_animals/reports_en.htm, accessed 19 Feb. 2011.

EC (2003a) (European Commission). *Third Report from the Commission to the Council and the European Parliament on the Statistics on the Number of Animals used for Experimental and other Scientific Purposes in the Member States of the European Union.* Brussels: EC. http://ec.europa.eu/environment/chemicals/lab_animals/reports_en.htm, accessed 19 Feb. 2011.

EC (2003b) (European Commission). IP/03/969. European Commission proposes strict ethical guidelines on EU funding of human embryonic stem cell research. www.tekno.dk/biosam/koordinering/PDFreferat/Koor_10092003-guidelines.pdf, accessed 13 Jun. 2010.

EC (2005) (European Commission). *Commission Staff Working Document. Annex to the Report on the Statistics on the Number of Animals used for Experimental and other Scientific Purposes in the Member States of the European Union in the Year 2002.* Brussels: EC. http://ec.europa.eu/environment/chemicals/lab_animals/reports_en.htm, accessed 19 Feb. 2011.

EC (2007) (European Commission). *Commission Staff Working Document. Annex to the Fifth Report on the Statistics on the Number of Animals used for Experimental and other Scientific Purposes in the Member States of the European Union.* Brussels: EC.http://ec.europa.eu/environment/chemicals/lab_animals/reports_en.htm, accessed 19 Feb. 2011.

EC (2010a) (European Commission). *Sixth Report on the Statistics on the Number of Animals used for Experimental and other Scientific Purposes in the Member States of the European Union.* Brussels: EC. http://ec.europa.eu/environment/chemicals/lab_animals/reports_en.htm, accessed 19 Feb. 2011.

EC (2010b) (European Commission). *Commission Staff Working Document Accompanying the Sixth Report on the Statistics on the Number of Animals used for Experimental and other Scientific Purposes in the Member States of the European Union.* Brussels: EC. http://ec.europa.eu/environment/chemicals/lab_animals/reports_en.htm, accessed 19 Feb. 2011.

ECHA (2008) (European Chemicals Agency). *Guidance on Registration.* Ver. 1.2. Helsinki: ECHA.

Eckert J (1997). Alternatives to animal experimentation in parasitology. *Vet Parasitol* 71(2–3): 99–120.

ECVAM (2010a) (European Centre for the Validation of Alternative Methods, Institute for Health and Consumer Protection, European Commission Joint Research Centre). About ECVAM. http://ecvam.jrc.it/, accessed 8 Jan. 2010.

ECVAM (2010b) (European Centre for the Validation of Alternative Methods, Institute for Health and Consumer Protection, European Commission Joint Research Centre). Method validation. http://ecvam.jrc.it/, accessed 4 Mar. 2011.

Edelman DB & Seth AK (2009). Animal consciousness: a synthetic approach. *Trends Neurosci* 32: 476–84.

Ehninger D & Kempermann G (2003). Regional effects of wheel running and environmental enrichment on cell genesis and microglia proliferation in the adult murine neocortex. *Cereb Cortex* 13: 845–51.

Ellaway R *et al.* (2005). The Edinburgh Electronic Veterinary Curriculum: an online program-wide learning and support environment for veterinary education. *J Vet Med Educ* 32(1): 38–46.

Ellinger-Ziegelbauer H *et al.* (2008). Prediction of a carcinogenic potential of rat hepatocarcinogens using toxicogenomics analysis of short-term *in vivo* studies. *Mutat Res* 637(1–2): 23–39.

Ennever FK & Lave LB (2003). Implications of the lack of accuracy of the lifetime rodent bioassay for predicting human carcinogenicity. *Regul Toxicol Pharmacol* 38: 52–7.

Ennever FK, Noonan TJ & Rosenkranz HS (1987). The predictivity of animal bioassays and short-term genotoxicity tests for carcinogenicity and non-carcinogenicity to humans. *Mutagenesis* 2: 73–8.

EPA (1991) (Environmental Protection Agency). *Report of the EPA Peer Review Workshop on Alpha2u-globulin: Association with Renal Toxicity and Neoplasia in the Male Rat.* Washington DC: EPA.

EPA (1999) (Environmental Protection Agency). *Guidelines for Carcinogen Risk Assessment.* Washington DC: Risk Assessment Forum, Environmental Protection Agency.

EPA (2003) (Environmental Protection Agency). *U.S. EPA's Process for IRIS Assessment Development and Review.* www.epa.gov/iris/process.htm, accessed 10 Dec. 2003.

EPA (2004a) (Environmental Protection Agency). Environmental Protection Agency. www.epa.gov, accessed 29 Jan. 2004.

EPA (2004b) (Environmental Protection Agency). What is IRIS? www.epa.gov/iris/intro.htm, accessed 29 Jan. 2004.

EPA (2004c) (Environmental Protection Agency). *IRIS Database for Risk Assessment.* www.epa.gov/iris/index.html, accessed 1 Jan. 2004.

EPA (2005) (Environmental Protection Agency). *Guidelines for Carcinogen Risk Assessment. EPA/630/P-03/001B.* Washington DC: Risk Assessment Forum, Environmental Protection Agency. www.epa.gov/cancerguidelines/, accessed 6 Feb. 2011.

EPA (2006) (Environmental Protection Agency). IRIS limitations. www.epa.gov/IRIS/limits.htm, accessed 5 Feb. 2006.

EPA (2010a) (Environmental Protection Agency). *High Production Volume (HPV) Challenge.* www.epa.gov/HPV/, accessed 25 Jun. 2010.

EPA (2010b) (Environmental Protection Agency). Oncologic™ – *A computer system to evaluate the carcinogenic potential of chemicals.* www.epa.gov/oppt/sf/pubs/oncologic.htm, accessed 16 Jun. 2010.

Epstein SS (1998). *The Politics of Cancer Revisited.* Fremont Center, NY: East Ridge Press.

Erickson HH & Clegg VL (1993). Active learning in cardiovascular physiology. In HI Modell and JA Michael (eds). *Promoting Active Learning in the Life Science Classroom. Ann N Y Acad Sci* 701: 107–8.

Eskola S *et al.* (1999). Environmental enrichment may alter the number of rats needed to achieve statistical significance. *Scand J Lab Anim Sci* 26: 134–44.

Ettlin RA & Prentice DE (2002). Unexpected tumour findings in lifetime rodent bioassay studies – what to do. *Toxicol Lett* 128: 17–33.

EU (2007) (European Union). *Treaty of Lisbon*. *Official J EU* 50: 1–272. http://europa.eu/lisbon_treaty/full_text/index_en.htm, accessed 24 Jun. 2010.

EU (2010) (European Union). Directive 2010/63/EU of the European Parliament and of the Council of 22 September 2010 on the protection of animals used for scientific purposes. *Official J EU* 276: 33–79.

Even M, Sandusky C & Barnard N (2006). Serum-free hybridoma culture: ethical, scientific and safety considerations. *Trends Biotechnol* 24(3): 105–8.

Evidence-Based Medicine Working Group (1992). Evidence-based medicine. A new approach to teaching the practice of medicine. *J Am Med Assoc* 286: 2420–5.

Farr S & Dunn RT (1999). Concise review: gene expression applied to toxicology. *Toxicol Sci* 50: 1–9.

Faverjon S *et al.* (2002). Beneficial effects of enriched environment following status epilepticus in immature rats. *Neurology* 59: 1356–64.

Fawver AL *et al.* (1990). A comparison of interactive videodisc instruction with live animal laboratories. *Am J Physiol* 259 (Advances in Physiology Education 4): 11–14.

FDA (2004) (Food and Drug Administration, US Department of Health and Human Services). *Innovation or Stagnation: Challenge and Opportunity on the Critical Path to New Medical Products*. www.fda.gov/ohrms/dockets/ac/04/briefing/2004-4052B1_11_ExecSum-Critical-Path.pdf, accessed 13 Jun. 2010.

Fearon R (2005). Cadaver donation programme boosts US veterinary education. *Vet Times* 35(20): 8–9.

Festing MFW (1997). Experimental design and husbandry. *Exp Gerontol* 32: 39–47.

Festing MFW (2004a). Is the use of animals in biomedical research still necessary in 2002? Unfortunately, 'Yes'. *Altern Lab Anim* 32(Suppl. 1B): 733–9.

Festing MFW (2004b). Good experimental design and statistics can save animals, but how can it be promoted? *Altern Lab Anim* 32(Suppl. 1A): 133–5.

Festing MFW & Altman DG (2002). Guidelines for the design and statistical analysis of experiments using laboratory animals. *ILAR J* 43: 244–57.

Festing MFW *et al.* (1998). Reducing the use of laboratory animals in biomedical research: problems and possible solutions. *Altern Lab Anim* 26, 283–301.

Fitzpatrick JL & Mellor DJ (2003). Survey of the views of graduates (1993 to 1997) on the undergraduate veterinary clinical curriculum in the British Isles. *Vet Rec* 153(13): 393–6.

Flecknell P (2008). Analgesia from a veterinary perspective. *Br J Anaesth* 101(1): 121–4.

Fleischer M (2007). Testing cost and testing capacity according to REACH requirements – results of a survey of independent and corporate GLP laboratories in the EU and Switzerland. *J Business Chem* 4: 96–114.

Fouts R (1995). Chimpanzee biomedical experiments: a question of efficacy. In *Poor Model Man: Experimenting on Chimpanzees: Proceedings of the First PACE (People Against Chimpanzee Experiments) Conference on the Use of Chimpanzees in Biomedical Research*. *Altern Lab Anim* 23: 584–91.

Fowler HS & Brosius EJ (1968). A research study on the values gained from dissection of animals in secondary school biology. *Sci Educ* 52(2): 55–7.

Francione G & Charlton A (1992). *Vivisection and Dissection in the Classroom: A Guide to Conscientious Objection*. Philadelphia, PA: Amer. Anti-Vivisection Soc.

Fry D (2004). Reduction by well-defined objectives. *Altern Lab Anim* 32(Suppl.1): 241–4.

Fung V, Barrett J & Huff J (1995). The carcinogenesis bioassay in perspective: application in identifying human hazards. *Environ Health Perspect* 103: 680–3.

Gad SC (1990). Model selection in toxicology: principles and practice. *J Am Coll Toxicol* 9: 291–302.

Gallas MM *et al.* (2005). Three-dimensional numerical simulation of dental implants as orthodontic anchorage. *Eur J Orthod* 27: 12–16.

Galle U & Bubna-Littitz H (1983). Modellzum Erlernen der Venenpunktionstechnik beim Hund. Eine Möglichkeit zur Reduzierung der Belastung lebender Versuchstiere im Studienbetrieb. [German]. Model for teaching venous puncture techniques in the dog. Possibility for reducing stress to living experimental animals in student teaching. *Zentralbl Veterinarmed A* 30(10): 796–9.

Gärtner K *et al.* (1980). Stress response of rats to handling and experimental procedures. *Lab Anim* 14: 267–74.

Genschow E *et al.* (2002). The ECVAM international validation study on *in vitro* embryotoxicity tests: results of the definitive phase and evaluation of prediction models. *Altern Lab Anim* 30(2): 151–76.

Genschow E *et al.* (2004). Validation of the embryonic stem cell test in the international ECVAM validation study on three *in vitro* embryotoxicity tests. *Altern Lab Anim* 32(3): 209–44.

Germani M *et al.* (2007). Evaluation of a basic physiologically based pharmacokinetic model for simulating the first-time-in-animal study. *Eur J Pharm Sci* 31(3–4): 190–201.

Gerner I *et al.* (2004). Development and prevalidation of a list of structure-activity relationship rules to be used in expert systems for prediction of the skin-sensitising properties of chemicals. *Altern Lab Anim* 32(5): 487–509.

Gerson LB & Van Dam J (2003). A prospective randomized trial comparing a virtual reality simulator to bedside teaching for training in sigmoidoscopy. *Endoscopy* 35(7): 569–575. Erratum in *Endoscopy* 2004; 36(2): 185.

Glazko G *et al.* (2005). Eighty percent of proteins are different between humans and chimpanzees. *Gene* 346: 215–19.

Gold LS & Zeiger E (eds) (2000). *Handbook of Carcinogenic Potency and Genotoxicity Databases*. Boca Raton, FL: CRC Press.

Gold LS, Slone TH & Ames BN (1998). What do animal cancer tests tell us about human cancer risk? Overview of analyses of the carcinogenic potency database. *Drug Metab Rev* 30: 359–404.

Gold LS *et al.* (1989). Interspecies extrapolation in carcinogenesis: prediction between rats and mice. *Environ Health Perspect* 81: 211–19.

Gold LS *et al.* (1999). Supplement to the Carcinogenic Potency Database (CPDB): results of animal bioassays published in the general literature in 1993 to 1994 and by the National toxicology Program in 1995 to 1996. *Environ Health Perspect* 107(Suppl. 4): 527–600.

Gold LS *et al.* (2005). Supplement to the Carcinogenic Potency Database (CPDB): results of animal bioassays published in the general literature through 1997 and by the National Toxicology Program in 1997–1998. *Toxicol Sci* 85: 747–808.

Gonder JC (2004). Resocialization and retirement of laboratory chimpanzees. *J Med Primatol* 33: 60–2.

Goodall J (1995). Why is it unethical to use chimpanzees in the laboratory? In *Poor Model Man: Experimenting on Chimpanzees: Proceedings of the First PACE (People Against Chimpanzee Experiments) Conference on the Use of Chimpanzees in Biomedical Research. Altern Lab Anim* 23: 615–20.

Goodall J (2003). Problems faced by wild and captive chimpanzees: finding solutions. In SJ Armstrong and RG Botzler (eds). *The Animal Ethics Reader.* London: Routledge.

Goodall J (2006). Ending research on non-human primates. *ALTEX* 23(Sp. Issue): 14–18.

Goodman JI (2001). A perspective on current and future uses of alternative models for carcinogenicity testing. *Toxicol Pathol* 29(Suppl. 1): 173–6.

Goodman S & Check E (2002). The great primate debate. *Nature* 417: 684–7.

Gottmann E *et al.* (2001). Data quality in predictive toxicology: reproducibility of rodent carcinogenicity experiments. *Environ Health Perspect* 109: 509–14.

Graham DJ *et al.* (2005). Risk of acute myocardial infarction and sudden cardiac death in patients treated with cyclo-oxygenase 2 selective and non-selective non-steroidal anti-inflammatory drugs: nested case-control study. *Lancet* 365: 475–81.

Greek CR & Greek JS (2000). *Sacred Cows and Golden Geese.* New York: Continuum.

Greek CR & Greek JS (2002a). *4th World Congress Point/Counterpoint: Is Animal Research Necessary in 2002?* Los Angeles: Americans for Medical Advancement.

Greek CR & Greek JS (2002b). *Specious Science.* New York: Continuum.

Green S & Goldberg AM (2004). TestSmart and toxic ignorance. *Altern Lab Anim* 32(Suppl. 1A): 359–63.

Greenfield CL *et al.* (1994). Integrating alternative models into the existing surgical curriculum. *J Vet Med Educ* 21(1): 23–7.

Greenfield CL *et al.* (1995). Comparison of surgical skills of veterinary students trained using models or live animals. *J Am Vet Med Assoc* 206(12): 1840–5.

Griffon DJ *et al.* (2000). Evaluation of a hemostasis model for teaching ovariohysterectomy in veterinary surgery. *Vet Surg* 29(4): 309–16.

Gruber FP & Dewhurst DG (2004). Alternatives to animal experimentation in biomedical education. *ALTEX* 21(Suppl 1): 33–48.

Gruber FP & Hartung T (2004). Alternatives to animal experimentation in basic research. *ALTEX* 21(Suppl. 1): 3–31.

Guengerich FP (2006). Cytochrome P450s and other enzymes in drug metabolism and toxicity. *AAPS J* 8(1): e101–11.

Guy JF & Frisby AJ (1992). Using interactive videodiscs to teach gross anatomy to undergraduates at Ohio State University. *Acad Med* 67: 132–3.

Hackam DG (2007). Translating animal research into clinical benefit: poor methodological standards in animal studies mean that positive results may not translate to the clinical domain. *Br Med J* 334: 163–4.

Hackam DG & Redelmeier DA (2006). Translation of research evidence from animals to humans. *J Am med Assoc* 296: 1731–2.

Hard GC & Whysner J (1994). Risk assessment of d-limonene: an example of male rat-specific renal tumorigens. *Crit Rev Toxicol* 24: 231–54.

Hart LA, Wood MW & Weng H (2005). Mainstreaming alternatives in veterinary medical education: resource development and curricular reform. *J Vet Med Educ* 32(4): 473–480.

Hartung T (2007). Food for thought... on cell culture. *ALTEX* 24(3): 143–7.

Hartung T (2008a). Food for thought... on animal tests. *ALTEX* 25(1): 3–9.

Hartung T (2008b). Thoughts on limitations of animal models. *Parkinsonism Relat Disord* 14(Suppl. 2): S81–3.

Haseman K (2000). Using the NTP database to assess the value of rodent carcinogenicity studies for determining human cancer risk. *Drug Metab Rev* 32: 169–86.

Hastwell PW *et al.* (2006). High-specificity and high-sensitivity genotoxicity assessment in a human cell line: validation of the GreenScreen® HC GADD45a-GFP genotoxicity assay. *Mutat Res* 607: 160–75.

Hawkins EC, Hansen B & Bunch BL (2003). Use of animation-enhanced video clips for teaching abnormal breathing patterns. *J Vet Med Educ* (Sp. Issue: Continuing Veterinary Education) 30(1): 73–7.

Hawkins P (2002). Recognizing and assessing pain, suffering and distress in laboratory animals: a survey of current practice in the UK with recommendations. *Lab Anim* 36(4): 378–95.

Heindl C, Hess A & Brune K (2008). Refinement and reduction in animal experimentation: options for new imaging techniques. *ALTEX* 25(2): 121–5.

Hellyer P *et al.* (1999). Attitudes of veterinary medical students, house officers, clinical faculty, and staff toward pain management in animals. *J Am Vet Med Assoc* 214(2): 238–44.

Hendriksen CFM (2006). Towards eliminating the use of animals for regulatory required vaccine quality control. *ALTEX* 23(3): 187–90.

Hendriksen CFM *et al.* (1987). The effects of reductions in the numbers of animals used for the potency assay of the diphtheria and tetanus components of adsorbed vaccines by the methods of the European pharmacopoeia. *J Biol Stand* 15: 353–62.

Hengstler JG *et al.* (1999). Interspecies differences in cancer susceptibility and toxicity. *Drug Metab Rev* 31: 917–70.

Henman MC & Leach GDH (1983). An alternative method for pharmacology laboratory class instruction using biovideograph videotape recordings. *Br J Pharmacol* 80: 591.

Higgins W (2008). San Marino bans all animal experiments. *Lifescape* 23: 14.

Hines SA *et al.* (2005). ATLES: the strategic application of Web-based technology to address learning objectives and enhance classroom discussion in a veterinary pathology course. *J Vet Med Educ* 32(1): 103–12.

Hoffmann S & Hartung T (2006). Toward an evidence-based toxicology. *Hum Exp Toxicol* 25: 497–513.

Holliday R (1996). Neoplastic transformation: the contrasting stability of human and mouse cells. *Cancer Surv* 28: 103–15.

Holmberg DL, Cockshutt JR & Basher AWP (1993). Use of a dog abdominal surrogate for teaching surgery. *J Vet Med Educ* 20(2): 61–2.

Holt RI *et al.* (2001). Computer assisted learning is an effective way of teaching endocrinology. *Clin Endocrinol (Oxf)* 55(4): 537–42.

Home Office (1998). *Report of the Animal Procedures Committee for 1997.* London: The Stationery Office.

Home Office (2010). *Statistics of Scientific Procedures on Living Animals: Great Britain 2009.* London: The Stationery Office. www.homeoffice.gov.uk/rds/scientific1.html, accessed 3 Nov. 2010.

Horn J *et al.* (2001). Nimodipine in animal model experiments of focal cerebral ischemia: a systematic review. *Stroke* 32: 2433–8.

Houdebine LM (2007). Transgenic animal models in biomedical research. *Methods Mol Biol* 360: 163–202.

Howe LM & Slater MR (1997). Student assessment of the educational benefits of a prepubertal gonadectomy program (preliminary findings). *J Vet Med Educ* 24(1): 12–17.

Howe LM, Boothe HW & Hartsfield SM (2005). Student assessment of the educational benefits of using a CD-ROM for instruction of basic surgical skills. *J Vet Med Educ* 32(1), 138–43.

HRA (2010) (Humane Research Australia). Statistics – Animal use in research & teaching, Australia. www.aahr.org.au/, accessed 13 Nov 2010.

HSUS (2007) (Humane Society of the United States). Swiss animal use statistics for 2005. *Pain Distress Rep* 7: 2.

HSUS & NEAVS (2008) (Humane Society of the United States & New England Anti-Vivisection Society). Federal bill introduced to end invasive research on chimpanzees. www.humanesociety.org/news/press_releases/2008/04/federal_bill_introduced_to_end_chimp_research_041708.html, accessed 6 Feb. 2011.

Huang SD & Aloi J (1991). The impact of using interactive video in teaching general biology. *Am Biol Teach* 53(5): 281–4.

Huff J (1999). Long-term chemical carcinogenesis bioassays predict human cancer hazards. Issues, controversies, and uncertainties. *Ann N Y Acad Sci* 895: 56–79.

Huff J (2002). Chemicals studied and evaluated in long-term carcinogenesis bioassays by both the Ramazzini Foundation and the National Toxicology Program. *Ann N Y Acad Sci* 982: 208–30.

Huggins J (2003). Alternatives to animal testing: research, trends, validation, regulatory acceptance. *ALTEX* 20(Suppl. 1): 3–61.

Hughes IE (2001). Do computer simulations of laboratory practicals meet learning needs? *Trends Pharmacol Sci* 22(2): 71–4.

IARC (1972–2004) (International Agency for Research on Cancer). *IARC Monographs on the Evaluation of Carcinogenic Risks to Humans. Volumes 1–82.* Lyon, France: IARC. http://monographs.iarc.fr/, accessed 29 Jun. 2010.

IARC (1999a) (International Agency for Research on Cancer). Objective and scope. www-cie.iarc.fr/monoeval/objectives.html, accessed 12 Jan. 2005.

IARC (1999b) (International Agency for Research on Cancer). *IARC Monographs Programme on the Evaluation of Carcinogenic Risks to Humans. Vol. 73: Some Chemicals that Cause Tumours of the Kidney or Urinary Bladder in Rodents, and Some Other Substances.* Lyon, France: IARC.

IARC (2005) (International Agency for Research on Cancer). *Internal Report 05/001: Report of the Advisory Group to Recommend Updates to the Preamble to the IARC Monographs.* Lyon, France: IARC.

IARC (2006a) (International Agency for Research on Cancer). Data for the Monographs. http://monographs.iarc.fr/ENG/Preamble/currenta4data0706. php, accessed 6 Feb. 2011.

IARC (2006b) (International Agency for Research on Cancer). Studies of cancer in experimental animals. http://monographs.iarc.fr/ENG/Preamble/current-b3studiesanimals0706.php, accessed 6 Feb. 2011.

ICCVAM (1997) (Interagency Coordinating Committee on the Validation of Alternative Methods). *Validation and Regulatory Acceptance of Toxicological Test Methods. A Report of the ad hoc Interagency Coordinating Committee on the Validation of Alternative Methods.* Research Triangle Park, NC: National Institute of Environmental Health Sciences.

ILAR (1992) (Institute for Laboratory Animal Research). *Recognition and Alleviation of Pain and Distress in Laboratory Animals.* Washington DC: National Academies Press.

ILAR (2011) (Institute for Laboratory Animal Research). *Guide for the Care and Use of Laboratory Animals* (8th edn). Washington DC: National Academies Press.

ISTCG (1997) (International Stroke Trial Collaborative Group). The International Stroke Trial (IST): a randomised trial of aspirin, subcutaneous heparin, or both, or neither, among 19,435 patients with acute ischaemic stroke. *Lancet* 349: 1569–81.

Itakura S (1994). Differentiated responses to different human conditions by chimpanzees. *Percept Mot Skills* 79(3 Pt 1):1288–90.

Izumi A & Kojima S (2004). Matching vocalizations to vocalizing faces in a chimpanzee (*Pan troglodytes*). *Anim Cogn* 7(3): 179–84.

Jefferies AR (2003). Curriculum development at Cambridge Veterinary School. *J Vet Med Educ* 30(3): 211–14.

Jennings P et al. (2004). Assessment of a new cell culture perfusion apparatus for *in vitro* chronic toxicity testing. Part 2: toxicological evaluation. *ALTEX* 21(2): 61–6.

Jensen K, Call J & Tomasello M (2007). Chimpanzees are rational maximizers in an ultimatum game. *Science* 318(5847): 107–9.

Jha A (2006, 3 Jun.). Questions raised over ban on research using great apes. *Guardian* www.guardian.co.uk/science/2006/jun/03/animalrights.research, accessed 15 Jun. 2010.

Johnson AL & Farmer JA (1989). Evaluation of traditional and alternative models in psychomotor laboratories for veterinary surgery. *J Vet Med Educ* 16(1): 11–14.

Johnson FM (1999). Carcinogenic chemical-response 'fingerprint' for male F344 rats exposed to a series of 195 chemicals: implications for predicting carcinogens with transgenic models. *Human Ecol Risk Assess* 5: 427–43.

Johnson FM (2001). Response to Tennant et al.: attempts to replace the NTP rodent bioassay with transgenic alternatives are unlikely to succeed. *Environ Mol Mutagen* 37: 89–92.

Jonas S et al. (2001). The failure of neuronal protective agents versus the success of thrombolysis in the treatment of ischemic stroke: the predictive value of animal models. *Ann N Y Acad Sci* 939: 257–67.

Jones NA, Olafson RP & Sutin J (1978). Evaluation of a gross anatomy program without dissection. *J Med Educ* 53: 198–205.

Josephson EM & Moore LJ (2006). An electronic instructor for gross anatomy dissection. *J Vet Med Educ* 33(3): 465–73.

Jukes N & Chiuia M (2003). *From Guinea Pig to Computer Mouse: Alternative Methods for a Progressive, Humane Education* (2nd edn). Leicester, UK: InterNICHE.

Kaminski J, Call J & Tomasello M (2008). Chimpanzees know, what others know, but not what they believe. *Cognition* 109: 224–34.

Kerckaert GA, LeBoeuf RA & Isfort RJ (1998). Assessing the predictiveness of the Syrian hamster embryo cell transformation assay for determining the rodent carcinogenic potential of single ring aromatic/nitroaromatic amine compounds. *Toxicol Sci* 41: 189–97. Erratum in *Toxicol Sci* 1998; 46: 420.

Kinzie MB, Strauss R & Foss J (1993). The effects of an interactive dissection simulation on the performance and achievement of high school biology students. *J Res Sci Teach* 30(8): 989–1000.

Kirkland D *et al.* (2005). Evaluation of the ability of a battery of three *in vitro* genotoxicity tests to discriminate rodent carcinogens and non-carcinogens I. Sensitivity, specificity and relative predictivity. *Mutat Res* 584: 1–256.

Kjellmer I (2002). Djurförsök är nödvändiga. Samordnade kontrollfunktioner låter sig svårligen studeras utan tillgång till naturens mest komplexa system: däggdjur och människa. [Swedish]. [Animal experiments are necessary. Coordinated control functions are difficult to study without the use of nature's most complex systems: mammals and human beings.] *Lakartidningen* 99: 1172–3.

Knight A (1999). Alternatives to harmful animal use in tertiary education. *Altern Lab Anim* 27(6): 967–74.

Knight A (2007a). Systematic reviews of animal experiments demonstrate poor human clinical and toxicological utility. *Altern Lab Anim* 35(6): 641–59.

Knight A (2007b). The poor contribution of chimpanzee experiments to biomedical progress. *J Appl Anim Welf Sci* 10(4): 281–308.

Knight A (2007c). The effectiveness of humane teaching methods in veterinary education. *ALTEX* 24(2): 91–109.

Knight A (2008a). 127 million non-human vertebrates used worldwide for scientific purposes in 2005. *Altern Lab Anim* 36(5): 494–6.

Knight A (2008b). Systematic reviews of animal experiments demonstrate poor contributions toward human healthcare. *Rev Recent Clin Trials* 3(2): 89–96.

Knight A (2008c). The beginning of the end for chimpanzee experiments? *Philos Ethics Humanit Med* 2008; 3: 16.

Knight A (2008d). Reviewing existing knowledge prior to conducting animal studies. *Altern Lab Anim* 36(6): 709–12.

Knight A (2008e). Non-animal methodologies within biomedical research and toxicity testing. *ALTEX* 25(3): 213–31.

Knight A (2008f). Advancing animal welfare standards within the veterinary profession. *REDVET: Revista electrónica de Veterinaria* 9(10B). A publication of Veterinaria.org.

Knight A (2010). Laboratory animal use in Great Britain in 2009. *AATEX* 15(2): 59–60.

Knight A, Bailey J & Balcombe J (2006a). Animal carcinogenicity studies: 1. Poor human predictivity. *Altern Lab Anim* 34(1): 19–27.

Knight A, Bailey J & Balcombe J (2006b). Animal carcinogenicity studies: 2. Obstacles to extrapolation of data to humans. *Altern Lab Anim* 34(1): 29–38.

Knight A, Bailey J & Balcombe J (2006c). Animal carcinogenicity studies: 3. alternatives to the bioassay. *Altern Lab Anim* 34(1): 39–48.

Knight A, Bailey J & Balcombe J (2006d). Animal carcinogenicity studies: implications for the REACH system. *Altern Lab Anim* 34(Suppl 1): 139–47.

Knight PA & Roberts PA (1987). An evaluation of some proposals for a reduction in the number of animals used for the potency testing of diphtheria and tetanus vaccines. *J Biol Stand* 15: 165–75.

Kojima S, Izumi A & Ceugniet M (2003). Identification of vocalizers by pant hoots, pant grunts and screams in a chimpanzee. *Primates* 44(3): 225–30.

Kolar R (2008, 17 Apr.) (German Animal Welfare Academy, Neubiberg). Personal communication to A Knight re: prevalence of invasive research on great apes within Germany.

Kopcha M *et al.* (2005). Practice-based education at Michigan State University. *J Vet Med Educ* 32(4): 555–61.

Koppanyi T & Avery MA (1966). Species differences and the clinical trial of new drugs: a review. *Clin Pharmacol Ther* 7: 250–70.

Krachun C *et al.* (2009). A competitive nonverbal false belief task for children and apes. *Dev Sci* 12(4): 521–35.

Kramer B *et al.* (2006). Getting it right: being smarter about clinical trials: a major NIH meeing led to recommendations for conducting better clinical trials. *PLoS Medicine* 3: 144.

Kramer JA *et al.* (2004). Acute molecular markers of rodent hepatic carcinogenesis identified by transcription profiling. *Chem Res Toxicol* 17: 463–70.

Kuhse H & Singer P (1998). *A Companion to Bioethics*. Malden, MA: Blackwell Publishing.

Kulpa-Eddy JA (2006). Overview of the regulatory requirements for the consideration of alternatives. *ALTEX* 23(Sp. Issue): 200–2.

Kulpa-Eddy J & Adams K (2008). U.S. perspective on the 'consideration of alternatives' regulatory requirement. *AATEX* 14(Sp. Issue: *Proc. 6th World Congress on Alternatives & Animal Use in the Life Sciences August 21–25, 2007, Tokyo, Japan*): 333–6.

Kumar AM *et al.* (2001). Client donation program for acquiring dogs and cats to teach veterinary gross anatomy. *J Vet Med Educ* 28(2): 73–7.

Kumar S *et al.* (2005). Placing confidence limits on the molecular age of the human-chimpanzee divergence. *Proc Natl Acad Sci USA* 102(52): 18842–7.

La Follette H & Shanks N (1994). Animal experimentation: the legacy of Claude Bernard. *Intnl Stud Philos Sci* 8: 195–210.

Langley G (2006). *Next of Kin: A Report on the Use of Primates in Experiments*. London: British Union for the Abolition of Vivisection.

Langley G *et al.* (2000). Volunteer studies replacing animal experiments in brain research: report and recommendations of a volunteers in research and testing workshop. *Altern Lab Anim* 28: 315–31.

Langley G *et al.* (2007). Replacing animal experiments: choices, chances and challenges. *Bioessays* 29(9): 918–26.

Langsch A & Nau H (2006). Metabolic activation for *in vitro* systems. *ALTEX* 23(Sp. Issue): 353–7.

Lavé T *et al.* (2007). Challenges and opportunities with modelling and simulation in drug discovery and drug development. *Xenobiotica* 37(10–11): 1295–1310.

Lazarou J & Pomeranz B (1998). Incidence of adverse drug reactions in hospitalized patients: a meta-analysis of prospective studies. *J Am Med Assoc* 279: 1200–5.

Lazzarini L *et al.* (2006). Experimental osteomyelitis: What have we learned from animal studies about the systemic treatment of osteomyelitis? *J Chemother* 18: 451–60.

Leake J (2007, 1 Mar.). Goodbye labs, hello chimp island heaven. *The Sunday Times*. www.timesonline.co.uk/tol/news/world/africa/article1596781.ece, accessed 6 Feb. 2011.

Leathard HL & Dewhurst DG (1995). Comparison of the cost effectiveness of a computer-assisted learning program with a tutored demonstration to teach intestinal motility to medical students. *Assoc Learn Technol J* 3(1): 118–25.

LeBoeuf RA *et al.* (1996). The pH 6.7 Syrian hamster embryo cell transformation assay for assessing the carcinogenic potential of chemicals. *Mutat Res* 356: 85–127.

Lee DS *et al.* (2003). Meta-analysis of the effects of endothelin receptor blockade on survival in experimental heart failure. *J Card Fail* 9: 368–74.

Lee KP *et al.* (2002). Association of journal quality indicators with methodological quality of clinical research articles. *J Amer Med Assoc,* 287: 2805–8.

Leib SL (2007). An *in vitro* model of central nervous system infection and regeneration: neuronal stem cells as targets of brain damage and regenerative therapies in bacterial meningitis. *ALTEX* 24(Sp. Issue): 90–2.

Leist M, Kadereit S & Schildknecht S (2008a). Food for thought … on the real success of 3R approaches. *ALTEX* 25(1): 17–24.

Leist M, Hartung T & Nicotera P (2008b). The dawning of a new age of toxicology. *ALTEX* 25(2): 103–14.

Leonard WH (1992). A comparison of student performance following instruction by interactive videodisc versus conventional laboratory. *J Res Sci Teach* 29(1): 93–102.

Levine ED, Mills DS & Houpt K (2005). Attitudes of veterinary students at one US college toward factors relating to farm animal welfare. *J Vet Med Educ* 32(4): 481–90.

Li AP (2005). Cell culture tool and method. U.S. Patent 20050101010.

Li AP (2008a). Human hepatocytes as an effective alternative experimental system for the evaluation of human drug properties: general concepts and assay procedures. *ALTEX* 25(1): 33–42.

Li AP (2008b). *In vitro* evaluation of human xenobiotic toxicity: scientific concepts and the novel integrated discrete multiple cell coculture (IdMOC) technology. *ALTEX* 25(1): 43–9.

Lichtenberg-Frate H *et al.* (2003). A yeast-based method for the detection of cyto- and genotoxicity. *Toxicol In Vitro* 17: 709–16.

Lieb MJ (1985). *Dissection: A Valuable Motivational Tool or a Trauma to the High School Student?* Evanston, Illinois: National College of Education. [Master Educ thesis].

Lilienblum W *et al.* (2008). Alternative methods to safety studies in experimental animals: role in the risk assessment of chemicals under the new European Chemicals legislation (REACH). *Arch Toxicol* 82(4): 211–36.

Lilienfield LS & Broering NC (1994). Computers as teachers: learning from animations. *Am J Physiol* 266(6 Pt 3): 47–54.

Lindl T, Völkel M & Kolar R (2005). Tierversuche in der biomedizinischen Forschung. Eine Bestandsaufnahme der klinischen Relevanz von genehmigten Tierversuchsvorhaben. [German]. [Animal experiments in biomedical research. An evaluation of the clinical relevance of approved animal experimental projects.] *ALTEX* 22(3): 143–51.

Linton A, Schoenfeld-Tacher R & Whalen LR (2005). Developing and implementing an assessment method to evaluate a virtual canine anatomy program. *J Vet Med Educ* 32(2): 249–54.

Lockhart DJ & Winzeler EA (2000). Genomics, gene expression and DNA arrays. *Nature* 405: 827–36.

Lois SW *et al.* (1991). The Carcinogenic Potency Database: analyses of 4000 chronic animal cancer experiments published in the general literature and by the U.S. National Cancer Institute/National Toxicology Program. *Environ Health Perspect* 96: 11–15.

Lucas C *et al.* (2002). Wound healing in cell studies and animal model experiments by Low Level Laser Therapy; were clinical studies justified? A systematic review. *Lasers Med Sci* 17: 110–34.

Luttun A & Verfaillie CM (2006). A perspective on stem cells as a tool for *in vitro* testing. *ALTEX* 23(Sp. Issue): 388–92.

Luy J (1998). *Tötungsfrage in der Tierschutzethik*. [German]. [*The Question of Killing in Animal Welfare Ethics*]. Berlin: Freie Universität Berlin. [Doctoral thesis].

Luy J (2007). Ethische und rechtliche Aspekte von Tierversuchen an Primaten. [German]. [Ethical and legal aspects of animal experiments on nonhuman primates]. *Dtsch Tierarztl Wochenschr* 114(3): 81–5.

Mackerle J (2004). Finite element modelling and simulations in dentistry: a bibliography 1990–2003. *Comput Methods Biomech Biomed Engin* 7: 277–303.

Macleod MR *et al.* (2005a). Systematic review and meta-analysis of the efficacy of melatonin in experimental stroke. *J Pineal Res* 38: 35–41.

Macleod MR *et al.* (2005b). Systematic review and meta-analysis of the efficacy of FK506 in experimental stroke. *J Cereb Blood Flow Metab* 25: 1–9.

MAP Coalition (2007) (Mandatory Alternatives Petition Coalition). *Petition to the US Food and Drug Administration for Mandatory Use of Non-Animal Methods in the Development and Approval of Drugs and Devices*. www.alternatives-petition. org/, accessed 16 Jun. 2010.

Mapstone J, Roberts I & Evans P (2003). Fluid resuscitation strategies: a systematic review of animal trials. *J Trauma* 55: 571–89.

Marshall JW *et al.* (2000). Clomethiazole protects against hemineglect in a primate model of stroke. *Brain Res Bull* 52: 21–9.

Matsuzawa T, Tomonaga M & Tanaka M (eds) (2006). *Cognitive Development in Chimpanzees*. Tokyo: Springer.

Matthews D (1998). Comparison of MacPig to fetal pig dissection in college biology. *Am Biol Teach* 60(3): 228–9.

Matthews EJ & Contrera JF (1998). A new highly specific method for predicting the carcinogenic potential of pharmaceuticals in rodents using enhanced MCASE QSAR-ES software. *Regul Toxicol Pharmacol* 28: 242–64.

Matthews RA (2008). Medical progress depends on animal models – doesn't it? *J R Soc Med* 101: 95–8.

Mauthe RJ *et al.* (2001). The Syrian hamster embryo (SHE) cell transformation assay: review of the methods and results. *Toxicol Pathol* 29(Suppl. 1): 138–46.

May SA (2003). The Royal Veterinary College, London. *J Vet Med Educ* 30(3): 215–17.

Mayer J (2007). Use of behavior analysis to recognize pain in small mammals. *Lab Anim (NY)* 36(6): 43–8.

McCaffrey S (1995). A medical student stands up for compassion. *Good Med* Autumn: 6–9.

McCollum TL (1987). *The Effect of Animal Dissections on Student Acquisition of Knowledge of and Attitudes Toward the Animals Dissected.* Cincinnati, OH: University of Cincinnati. [Doctoral thesis].

McGrew WC (1994). Cultural implications of differences between populations of free ranging chimpanzees in Africa. In RA Gardner *et al.* (eds). *The Ethological Roots of Culture.* Dordrecht, The Netherlands: Kluwer Academic. 61–79.

Mead R (1988). *The Design of Experiments.* New York: Cambridge University Press.

Medical Research Council (1993). *Responsibility in the Use of Animals in Medical Research.* London: MRC.

Meijers JM, Swaen GM & Bloemen LJ (1997). The predictive value of animal data in human cancer risk assessment. *Regul Toxicol Pharmacol* 25: 94–102.

Meyers NM (1983). Government regulation of nonhuman primate facilities. *J Med Primatol* 12(4): 169–83.

Mill JS. (1971). *Essential Works of John Stuart Mill* (4th edn). New York: Bantam Books.

Mitterhauser M & Toegel S (2008). An *in vitro* model for the comparative evaluation of bone seeking pharmaceuticals. *ALTEX* 25(1): 51–5.

Modell JH *et al.* (2002). Using the human patient simulator to educate students of veterinary medicine. *J Vet Med Educ* 29(2): 111–16.

Moher D, Schulz KF & Altman DG (2001). The CONSORT statement: revised recommendations for improving the quality of reports of parallel-group randomised trials. *Lancet* 357: 1191–4.

Monro A (1993a). How useful are chronic (life-span) toxicology studies in rodents in identifying pharmaceuticals that pose a carcinogenic risk to humans? *Adverse Drug React Toxicol Rev* 12: 5–34.

Monro A (1993b). The paradoxical lack of inter-species correlations between plasma concentrations and chemical carcinogenicity. *Regul Toxicol Pharmacol* 18: 115–35.

Monro A (1996). Are lifespan rodent carcinogenicity studies defensible for pharmaceutical agents? *Exp Toxicol Pathol* 48: 155–66.

Monro A & Davies TS (1993). High dose levels are not necessary in rodent studies to detect human carcinogens. *Cancer Lett* 75: 183–94.

Monro AM & MacDonald JS (1998). Evaluation of the carcinogenic potential of pharmaceuticals. Opportunities arising from the International Conference on Harmonisation. *Drug Safety* 18: 309–19.

Monro A & Mordenti J (1995). Expression of exposure in negative carcinogenicity studies: dose/body weight, dose/body surface area, or plasma concentrations? *Toxicol Pathol* 23: 187–98.

Montag T *et al.* (2007). Safety testing of cell-based medicinal products: opportunities for the monocyte activation test for pyrogens. *ALTEX* 24(2): 81–9.

More D & Ralph CL (1992). A test of effectiveness of courseware in a college biology class. *J Educ Technol Syst* 21: 79–84.

Mori T *et al.* (2006). [Evaluation of a dog abdominal surrogate model for teaching basic surgical skills by veterinary students]. [Japanese]. *J Jap Vet Med Assoc* 59(2): 122–5.

Morton DB (2000). Self-consciousness and animal suffering. *Biologist* 47(2): 77–80.

Moynihan J *et al.* (1990). The effects of handling on antibody production, mitogen responses, spleen cell number, and lymphocyte subpopulations. *Life Sci* 46: 1937–44.

Murthy B (2007). Relevance of *in vitro* toxicology studies in risk assessment. *ALTEX* 24(3): 174–7.

Nahmias YK, Gao BZ & Odde DJ (2004). Dimensionless parameters for the design of optical traps and laser guidance systems. *Appl Opt* 43, 3999–4006.

NAVS (2007) (National Anti-Vivisection Society [UK]). Declaration calls for European primate test ban. www.navs.org.uk/take_action/39/0/812/, accessed 13 Jun. 2010.

NCI (1978) (National Cancer Institute). *Bioassay for Aniline Hydrochloride for Possible Carcinogenicity. CAS No. 142-04-1. ITS Carcinogenesis Technical Report Series No. 130. Department of Health, Education and Welfare (DHEW) Publication No. NIH 78-1385*. Bethesda, MD: DHEW.

Neugebauer EA (2009). 2. Evidence-based medicine – a possible model for evidence-based toxicology? 2.1 translation of evidence-based medicine into practice. *Hum Exp Toxicol* 28(2–3): 105–7.

Nevalainen T *et al.* (1999). FELASA guidelines for education of specialists in laboratory animal science (Category D). *Lab Anim* 33: 1–15.

NHMRC (2003) (National Health and Medical Research Council). *Policy on the Care and Use of Non-Human Primates for Scientific Purposes*. Canberra: NHMRC.

NHMRC (2004) (National Health and Medical Research Council). *Australian Code of Practice for the Care and Use of Animals for Scientific Purposes* (7th edn). Canberra: Aust. Govt Publish. Svce.

Nicol C (2010). The progress we've made in assessing animal interests. *AWSELVA J* 14(2): 6–8.

NIH (2006) (National Institutes of Health). Understanding clinical trials. http://clinicaltrials.gov/ct2/info/understand, accessed 17 Jun. 2010.

NIND & SSG (1995) (The National Institute of Neurological Disorders & Stroke rt-PA Stroke Study Group). Tissue plasminogen activator for acute ischemic stroke. *N Engl J Med* 333: 1581–8.

NRC (2007) (National Research Council). *Toxicity Testing in the 21st Century: A Vision and a Strategy*. Washington DC: National Academies Press.

NTP (2002) (National Toxicology Program). *National Toxicology Program Report on Carcinogens* (10th edn). http://ntp.niehs.nih.gov/ntpweb/index.cfm?objectid=72016262-BDB7CEBA-FA60E922B18C2540, accessed 13 Jan. 2005.

NTP (2004) (National Toxicology Program). *A National Toxicology Program for the 21st Century: A Roadmap for the Future*. Research Triangle Park, NC: NTP, National Institute of Environmental Health Sciences.

Nuffield Council on Bioethics (2005). *The Ethics of Research Involving Animals*. London: Nuffield Council on Bioethics.

O'Collins VE *et al.* (2006). 1026 experimental treatments in acute stroke. *Ann Neurol* 59: 467–77.

O'Connor AM (1997). Barriers to regulatory acceptance. In LFM Van Zutphen and M Balls (eds). *Animal Alternatives, Welfare and Ethics*. Amsterdam: Elsevier Science BV. 1173–6.

Odde DJ & Renn MJ (2000). Laser-guided direct writing of living cells. *Biotechnol Bioeng* 67: 312–18.

OECD (2003) (Organisation for Economic Cooperation and Development). *OECD Series on Testing and Assessment: No. 34: Guidance Document on the Validation and International Acceptance of New and Updated Test Methods for Hazard Assessment*. Paris: OECD.

Of Human and Non-Human Animals (2007). San Marino bans vivisection. http://globalphilosophy.blogspot.com/2007/10/san-marino-bans-vivisection. html, accessed 13 Jun. 2010.

Oktem IS *et al.* (2000). Therapeutic effect of tirilazad mesylate (U-74006F), mannitol, and their combination, on experimental ischemia. *Res Exp Med* 199: 231–42.

OLAW (2002) (Office of Laboratory Animal Welfare, National Institutes of Health). *Public Health Service Policy on Humane Care and Use of Laboratory Animals*. Bethesda, MD: OLAW.

Olsen D *et al.* (1996). Evaluation of a hemostasis model for teaching basic surgical skills. *Vet Surg* 25(1): 49–58.

Olson H *et al.* (1998). The predictivity of the toxicity of pharmaceuticals in humans from animal data – an interim assessment. *Toxicol Lett* 102–3, 535–8.

Onishi KH & Baillargeon R (2005). Do 15-month-old infants understand false beliefs? *Science* 308: 255–8.

Orlans FB (1998). History and ethical regulation of animal experimentation: an international perspective. In H Kuhse and P Singer (eds). *A Companion to Bioethics*. Oxford: Blackwell. 399–410.

Orlans FB (2000). Research on animals, law, legislative, and welfare issues in the use of animals for genetic engineering and xenotransplantation. In TH Murray and MJ Mehlman (eds). *Encyclopaedia of Ethical, Legal and Political Issues in Biotechnology*. Hoboken, NJ: John Wiley & Sons. 1020–30.

Orsière T *et al.* (2006). Genotoxic risk assessment of pathology and anatomy laboratory workers exposed to formaldehyde by use of personal air sampling and analysis of DNA damage in peripheral lymphocytes. *Mutat Res* 605: 30–41.

Osswald W (1992). Etica da investigação no animal e aplicação ao homem. [Portugese]. [Ethics of animal research and application to humans.] *Acta Medica Portuguesa* 5: 222–5.

Paparella M *et al.* (2002). The use of quantitative image analysis in the assessment of *in vitro* embryotoxicity endpoints based on a novel embryonic stem cell clone with endoderm-related GFP expression. *Toxicol In Vitro* 16(5): 589–97.

Parr LA (2003). The discrimination of faces and their emotional content by chimpanzees (*Pan troglodytes*). *Ann N Y Acad Sci* 1000: 56–78.

Participatory Politics and Sunlight Foundations (2008). S.1916 Chimp Haven is Home Act. www.opencongress.org/bill/110-s1916/show, accessed 21 Jun. 2010.

Passineau MJ, Green EJ & Dietrich WD (2001). Therapeutic effects of environmental enrichment on cognitive function and tissue integrity following severe traumatic brain injury in rats. *Exp Neurol* 168: 373–84.

Patronek GJ & Rauch A (2007). Systematic review of comparative studies examining alternatives to the harmful use of animals in biomedical education. *J Amer Vet Med Assoc* 230(1): 37–43.

Paul E & Podberscek A (2000). Veterinary education and students' attitudes towards animal welfare. *Vet Rec* 146(10): 269–72.

Pavletic MM *et al.* (1994). An assessment of the outcome of the alternative medical and surgical laboratory program at Tufts University. *J Amer Vet Med Assoc* 205(1): 97–100.

Pawlik WW (1998). Znaczenie zwierzat w badaniach biomedycznych. [Polish]. [The significance of animals in biomedical research.] *Folia Medica Cracoviensia* 39: 175–82.

Paylor R *et al.* (1992). Brief exposure to an enriched environment improves performance on the Morris water task and increases hippocampal cytosolic protein kinase C activity in young rats. *Behav Brain Res* 52: 49–59.

Pellizzer C, Bremer S & Hartung T (2005). Developmental toxicity testing from animal towards embryonic stem cells. *ALTEX* 22(2): 47–57.

Pellizzer C *et al.* (2004a). Monitoring of teratogenic effects *in vitro* by analysing a selected gene expression pattern. *Toxicol In Vitro* 18(3): 325–35.

Pellizzer C *et al.* (2004b). Detection of tissue specific effects by methotrexate on differentiating mouse embryonic stem cells. *Birth Defects Res B Dev Reprod Toxicol* 71(5): 331–41.

Pereira S & Tettamanti M (2005). Ahimsa and alternatives – the concept of the 4th R. The CPCSEA in India. *ALTEX* 22(1): 3–6.

Perel P *et al.* (2007). Comparison of treatment effects between animal experiments and clinical trials: systematic review. *Br Med J* 334: 197–200.

Peters TS (2005). Do preclinical testing strategies help predict human hepatotoxic potentials? *Toxicol Pathol* 33: 146–54.

Peto R *et al.* (1984). The TD50: A proposed general convention for the numerical description of the carcinogenic potency of chemicals in chronic-exposure animal experiments. *Environ Health Perspect* 58: 1–8.

Phelps JL, Nilsestuen JO & Hosemann S (1992). Assessment of effectiveness of videodisc replacement of a live animal physiology laboratory. *Distinguished Papers Monograph, Am Assoc Resp Care.*

Phillips CJC (2005). Meta-analysis – a systematic and quantitative review of animal experiments to maximise the information derived. *Anim Welf* 14: 333–8.

Pichler WJ (2007). Predicting drug hypersensitivity by *in vitro* tests. *ALTEX* 24(Sp. Issue): 49–52.

Pienta RJ, Poiley JA & Lebherz WB 3rd (1977). Morphological transformation of early passage golden Syrian hamster embryo cells derived from cryopreserved primary cultures as a reliable *in vitro* bioassay for identifying diverse carcinogens. *Int J Cancer* 19: 642–55.

Pinkney RD *et al.* (2001). Impact of a computer-based auto-tutorial program on parasitology test scores of four consecutive classes of veterinary medical students. *J Vet Med Educ* 28(3): 136–9.

Pippin JJ (2008). 'MAP' for improving drug testing: Mandatory Alternatives Petition urges FDA to require use of replacements to animal testing. *Genet Eng Biotech News* 28(5).

Poignet H, Nowicki JP & Scatton B (1992). Lack of neuroprotective effect of some sigma ligands in a model of focal cerebral ischemia in the mouse. *Brain Res* 596: 320–4.

Postle M *et al.* (2003). *Assessment of the Impact of the New Chemicals Policy on Occupational Health: Final Report: Prepared for European Commission – Environment Directorate-General.* Loddon, Norfolk: Risk & Policy Analysts Limited.

Pound P *et al.* (2004). Where is the evidence that animal research benefits humans? *Br Med J* 328: 514–17.

Premack D & Woodruff G (1978). Does the chimpanzee have a theory of mind? *Behav Brain Sci* 4: 515–26.

Prentice ED *et al.* (1977). Stereoscopic anatomy: evaluation of a new teaching system in human gross anatomy. *J Med Educ* 52: 758–63.

Project R&R (2009). International bans. www.releasechimps.org/mission/end-chimpanzee-research/country-bans/, accessed 21 Jun. 2010.

Purcell WM & Atterwill CK (1995). Mast cells in neuroimmune function: neurotoxicological and neuropharmacological perspectives. *Neurochem Res* 20: 521–32.

Rall DP (2000). Laboratory animal tests and human cancer. *Drug Metab Rev* 2: 119–28.

RCVS (2010) (Royal College of Veterinary Surgeons). *Guide to Professional Conduct.* London: RCVS.

Reinhardt V & Reinhardt A (2006). *Variables, Refinement and Environmental Enrichment for Rodents and Rabbits Kept in Research Institutions.* Washington, DC: Animal Welfare Institute.

Repubblica di San Marino (2007). *Noi Capitani Reggenti la Serenissima Repubblica di San Marino.* [Italian].

Richardson EF, Gregory CR & Sucre E (1994). Enhancement of the surgical education of fourth year veterinary students by participation in juvenile ovario-hysterectomy and castration program. *Vet Surg* 23(5): 415.

Roberts I *et al.* (2002). Does animal experimentation inform human healthcare? Observations from a systematic review of international animal experiments on fluid resuscitation. *Br Med J* 324: 474–6.

Rochon PA *et al.* (1994). Evaluating the quality of articles published in journal supplements compared with the quality of those published in the parent journal. *J Amer Med Assoc* 272: 108–13.

Rogers DA *et al.* (1998). Computer-assisted learning versus a lecture and feedback seminar for teaching a basic surgical technical skill. *Amer J Surg* 175(6): 508–10.

Rose T (2007, 2 Aug.). Going ape over human rights. CBC News. www.cbc.ca/news/viewpoint/vp_rose/20070802.html, accessed 21 Jun. 2010.

Ross-Degnan D *et al.* (1993). Examining product risk in context. Market withdrawal of zomepirac as a case study. *J Am Med Assoc* 270: 1937–42.

Rowan A (1993). Replacement alternatives and the concept of alternatives. In AM Goldberg, LFM Van Zutphen and ML Principe (eds). *The World Congress on Alternatives and Animal Use in the Life Sciences: Education, Research, Testing (Alternative Methods in Toxicology).* New Rochelle, NY: Mary Ann Liebert. 1–10.

Rudas P (1993). Hypermedia in veterinary education. In G Hencsey and G Renner (eds). *Proceedings of the 3rd Annual International Conference and Exhibition on CAD/CAM/CAE/CIM. Applications for Manufacturing and Productivity.* Budapest: World Comput. Graphics Assoc. 212–18.

Rusche B (2003). The 3Rs and animal welfare – conflict or the way forward? *ALTEX* 20(Suppl. 1): 63–76.

Russell WMS & Burch RL (1959). *The Principles of Humane Experimental Technique.* London: Methuen.

Ruvolo M (2004). Comparative primate genomics: the year of the chimpanzee. *Curr Opin Genet Dev* 14: 650–6.

Sager M (2006). Use of analgesics in experiments. *ALTEX* 23(Suppl): 103–9.

Samsel RW *et al.* (1994). Cardiovascular physiology teaching: computer simulations vs. animal demonstrations. *Adv Physiol Educ* 11: 36–46.

Sandusky C *et al.* (2006). Strategies to reduce animal testing in US EPA's HPV program. *ALTEX* 23(Sp. Issue): 150–2.

Sauer UG, Kolar R & Rusche B (2005). Die Verwendung transgener Tiere in der biomedizinischen Forschung in Deutschland. Teil 1: Sachstandsbericht 2001–2003. [German]. [The use of transgenic animals in biomedical research in Germany. Part 1: Status Report 2001–2003.] *ALTEX* 22: 233–46.

Sauer UG, Kolar R & Rusche B (2006). Die Verwendung transgener Tiere in der biomedizinischen Forschung in Deutschland. Teil 2: Ethische Bewertung der Verwendung transgener Tiere in der biomedizinischen Forschung und Perspektiven fur die Umstellung der Forschung auf tierversuchsfreie Verfahren. [German]. [The use of transgenic animals in biomedical research in Germany. Part 2: Ethical evaluation of the use of transgenic animals in biomedical research and perspectives for the changeover in research to research animal-free methods.] *ALTEX* 23: 3–16.

Scalese RJ & Issenberg SB (2005). Effective use of simulations for the teaching and acquisition of veterinary professional and clinical skills. *J Vet Med Educ* 32(4): 461–7.

Schade R *et al.* (2005). Chicken egg yolk antibodies (Ig-Y-technology): a review of progress in production and use in research and human and veterinary medicine. *Altern Lab Anim* 33: 129–54.

Scheld WM (1987). Therapy of streptococcal endocarditis: correlation of animal model and clinical studies. *J Antimicrob Chemother* 20(Suppl. A): 71–85.

Schindler S *et al.* (2006a). International validation of pyrogen tests based on cryopreserved human primary blood cells. *J Immunol Methods* 316(1–2): 42–51.

Schindler S *et al.* (2006b). Pyrogen testing of lipidic parenterals with a novel *in vitro* test: application of the IPT based on cryopreserved human whole blood. *Pharmeur Sci Notes* 1: 1–7.

Schindler S *et al.* (2007). Fever in the test tube: towards a human(e) pyrogen test. *ALTEX* 24(Sp. Issue): 60–2.

Schoeffner DJ & Thorgeirsson UP (2000). Susceptibility of nonhuman primates to carcinogens of human relevance. *In Vivo* 14: 149–56.

Schrattenholz A & Klemm M (2007). Neuronal cell culture from human embryonic stem cells as *in vitro* model for neuroprotection. *ALTEX* 24(1): 9–15.

Schulz KF (2005). Assessing allocation concealment and blinding in randomised controlled trials: why bother? *Equine Vet J* 37: 394–5.

Schuppli CA & Fraser D (2005). The interpretation and application of the three Rs by animal ethics committee members. *Altern Lab Anim* 33(5): 487–500.

Scialli AR (2008). The challenge of reproductive and developmental toxicology under REACH. *Regul Toxicol Pharmacol* 51(2): 244–50.

Scopus (2006). Scopus in detail: what does it cover? www.info.scopus.com/detail/what/, accessed 1 Mar. 2007.

Seiler A *et al.* (2002). Improving the embryonic stem cell test (EST) by establishing molecular endpoints of tissue specific development using murine embryonic stem cells (D3 cells). *ALTEX* 19(Suppl. 1): 55–63.

Seiler A *et al.* (2004). Improvement of an *in vitro* stem cell assay for developmental toxicity: the use of molecular endpoints in the embryonic stem cell test. *Reprod Toxicol* 18(2): 231–40.

Self DJ, Pierce AB & Shadduck J (1994). A survey of the teaching of ethics in veterinary education. *J Am Vet Med Assoc* 204(6): 944–5.

Self DJ *et al.* (1991). Study of the influence of veterinary medical education on the moral development of veterinary students. *J Am Vet Med Assoc* 198(5): 782–7.

Self DJ *et al.* (1996). Clarifying the relationship of veterinary medical education and moral development. *J Am Vet Med Assoc* 209(12): 2002–4.

Serpell JA (2005). Factors influencing veterinary students' career choices and attitudes to animals. *J Vet Med Educ* 32(4): 491–6.

Shirai T *et al.* (1984). Effects of butylated hydroxyanisole, butylated hydroxytoluene, and NaCl on gastric carcinogenesis initiated with N-methyl-N-nitro-N-nitrosoguanidine in F344 rats. *J Natl Cancer Inst* 72: 1189–98.

Siegford JM *et al.* (2005). Integrating animal welfare into veterinary education: using an online, interactive course. *J Vet Med Educ* 32(4): 497–504.

Silk JB *et al.* (2005). Chimpanzees are indifferent to the welfare of unrelated group members. *Nature* 437(7063): 1357–9.

Silliman CC & Wang M (2006). The merits of *in vitro* versus *in vivo* modelling in investigation of the immune system. *Environ Toxicol Pharmacol* 21(2): 123–34.

Silva RMG, Matera JM & Ribeiro AACM (2004). Preservation of cadavers for surgical technique training. *Vet Surg* 33(6): 606–8.

Simpson RM & Meuten DJ (1992). Development of a teaching laboratory aid for instruction of fine needle aspiration biopsy cytology technique. *Vet Clin Path* 21(2): 40–4.

Singer P (1990). *Animal Liberation: A New Ethics for our Treatment of Animals* (2nd edn). New York: New York Review/Random House.

Smeak DD *et al.* (1991). Evaluation of video tape and a simulator for instruction of basic surgical skills. *Vet Surg* 20(1): 30–6.

Smeak DD *et al.* (1994). Evaluation of an autotutorial-simulator program for instruction of hollow organ closure. *Vet Surg* 23(6): 519–28.

Smith BP & Walsh DA (2003). Teaching the art of clinical practice: the veterinary medical teaching hospital, private practice, and other externships. *J Vet Med Educ* 30(3): 203–6.

Smith JA & Boyd KM (eds) (2002). *The Boyd Group Papers on the Use of Non-Human Primates in Research and Testing*. Leicester, UK: The British Psychological Society.

Specht PC (1988). Computer graphics interface to a complex simulation. *P R Health Sci J* 7(2): 184–8.

Stafford N (2007, 26 Apr.). Chimp denied a legal guardian. *Nature News*. www.nature.com/news/2007/070423/full/news070423-9.html, accessed 22 Jun. 2010.

STAIR (1999) (Stroke Therapy Academic Industry Roundtable). Recommendations for standards regarding preclinical neuroprotective and restorative drug development. *Stroke* 30: 2752–8.

State of California (2000). *Civil Code Section 1833-1840*. www.leginfo.ca.gov/cgi-bin/displaycode?section=civ&group=01001-02000&file=1833-1840, accessed 23 Jun. 2010.

State of New Jersey (2007). *An Act Concerning the Use of Animals in Product Testing and Supplementing Title 4 of the Revised Statutes*. Chapter 210. www.njleg.state. nj.us/2006/Bills/PL07/210_.HTM, accessed 23 Jun. 2010.

State of New York (2008). *New York Public Health Law. Chapter 45. Of the Consolidated Laws. Article 5. Laboratories. Title I. General Provisions: State Laboratories; Approved Laboratories*. www.aavs.org/images/NewYorkLaw.pdf, accessed 23 Jun. 2010.

Stephens ML, Alvino GM & Branson JB (2002). Animal pain and distress in vaccine testing in the United States. *Dev Biol (Basel)* 111: 213–16.

Stephens ML *et al*. (1998). Unrelieved pain and distress in animals: an analysis of USDA data on experimental procedures. *J Appl Anim Welf Sci* 1: 15–26.

Strauss RT & Kinzie MB (1994). Student achievement and attitudes in a pilot study comparing an interactive videodisc simulation to conventional dissection. *Am Biol Teach* 56(7): 398–402.

Stull L (2000). Illinois vet school survey – student quotes re: physiology labs. [unpublished]. www.humanelearning.info/resources/surveys.htm, accessed 25 Jun. 2010.

Sturma D (1999). Person. In HJ Sandkühler (ed.). *Enzyklopädie Philosophie* (2002 CD-ROM edn). Hamburg: Meiner.

Sutherland LM *et al*. (2006). Surgical simulation: a systematic review. *Ann Surg* 243(3): 291–300.

Takayanagi M *et al*. (2007). [Formaldehyde concentrations in the breathing zone of medical students during gross anatomy laboratory in Toho University.] [Japanese]. *Kaibogaku Zasshi* 82(2): 45–51.

Taylor K *et al*. (2008). Estimates for worldwide laboratory animal use in 2005. *Altern Lab Anim* 36: 327–42.

Taylor R (2001). A step at a time: New Zealand's progress toward hominid rights. *Anim Law* 7(35): 35–43.

TeGenero (2006a). Frequently asked questions regarding TGN1412. www.tegenero.com/news/faqs_re_tgn1412/index.php, accessed 18 Apr. 2006.

TeGenero (2006b). Statement re: TGN1412. www.tegenero.com/news/statement_re_tgn1412/index.php, accessed 18 Apr. 2006.

Tennant RW *et al*. (1990). Prediction of the outcome of rodent carcinogenicity bioassays currently being conducted on 44 chemicals by the National toxicology Program. *Mutagenesis* 5: 3–14.

Thon *et al*. (2002). Welfare evaluation of genetically modified mice – An inventory of reports to the Danish Animal Experiments Inspectorate. *Scand J Lab Anim Sci* 29: 45–53.

Times Newspapers (2007). Goodbye labs, hello chimp island heaven. *The Sunday Times*. www.timesonline.co.uk/tol/news/world/africa/article1596781.ece, accessed 13 Jun. 2010.

Tomatis L & Wilbourn J (1993). Evaluation of carcinogenic risk to humans: the experience of IARC. In O Iversen (ed.). *New Frontiers in Cancer Causation*. Washington DC: Taylor and Francis. 371–87.

Tomatis L *et al.* (1989). Human carcinogens so far identified. *Jap J Cancer Res* 80: 795–807.

Uchiyama I. (2010). [Toxicity of formaldehyde exposure and the details of its control measures.] [Japanese]. *Kaibogaku Zasshi* 85(1): 29–34.

Ullrich A *et al.* (2007). Use of a standardised and validated long-term human hepatocyte culture system for repetitive analyses of drugs: repeated administrations of acetaminophen reduces albumin and urea secretion. *ALTEX* 24(1): 35–40.

US Government Printing Office (1985). *The Animal Welfare Act – As Amended. Known as the 'Improved Standards for Laboratory Animal' 7 US Code §§2131-2159. Public Law 99-198 7 Code of Federal Regulations 2.22, 2.80, and 371.2 (g).* Washington DC: U.S. Government Printing Office.

USDA (2000) (US Department of Agriculture, Animal and Plant Health Inspection Service, Animal Care). *Animal Care Survey: USDA Employee Survey on the Effectiveness of IACUC Regulations.* Riverdale, MD: USDA.

USDA (2005) (US Dept of Agriculture, Animal and Plant Health Inspection Service, Animal Care). *FY 2005 AWA Inspections.* www.aphis.usda.gov/animal_welfare/downloads/awreports/awreport2005.pdf, accessed 10 Mar. 2010.

USDA (2011a) (US Dept of Agriculture, Animal and Plant Health Inspection Service, Animal Care). Policy #11: painful and distressful procedures. In *Animal Care Policy Manual.* www.aphis.usda.gov/animal_welfare/policy.php, accessed 21 Apr. 2011.

USDA (2011b) (US Dept of Agriculture, Animal and Plant Health Inspection Service, Animal Care). Policy #12: consideration of alternatives to painful/distressful procedures. In *Animal Care Policy Manual.* www.aphis.usda.gov/animal_welfare/policy.php, accessed 21 Apr. 2011.

Van der Valk J *et al.* (1999). Alternatives to the use of animals in higher education. The report and recommendations of ECVAM workshop 33. *Altern Lab Anim* 27: 39–52.

Van der Worp HB *et al.* (2005). Methodological quality of animal studies on neuroprotection in focal cerebral ischaemia. *J Neurol* 252: 1108–14.

Van Loo PLP *et al.* (2001). Do male mice prefer or avoid each other's company? Influence of hierarchy, kinship, and familiarity. *J Appl Anim Welf Sci* 4: 91–103.

Van Wilgenburg H, Van Schaick Zillesen PG & Krulichova I (2003). Sample power and ExpDesign: tools for improving design of animal experiments. *Lab Anim* 32: 39–43.

Van Wilgenburg H, Van Schaick Zillesen PG & Krulichova I (2004). Experimental design: computer simulation for improving the precision of an experiment. *Altern Lab Anim* 32(Suppl. 1B): 607–11.

Van Zutphen LFM (2001). Introduction. In LFM Van Zutphen, V Baumans and C Beynen (eds). *Principles of Laboratory Animal Science.* Amsterdam: Elsevier. 2–5.

VandeBerg JL *et al.* (2005). A unique biomedical resource at risk. *Nature* 437: 30–2.

Varki A & Altheide TK (2005). Comparing the human and chimpanzee genomes: searching for needles in a haystack. *Genome Res* 15(12): 1746–58.

Vedani A, Dobler M & Lill MA (2005). Virtual test kits for predicting harmful effects triggered by drugs and chemicals mediated by specific proteins. *ALTEX* 22(3): 123–34.

Vedani A, Lill MA & Dobler M (2007a). Predicting the toxic potential of drugs and chemicals *in silico*. *ALTEX* 24(Sp. Issue): 63–6.

Vedani A *et al.* (2007b). Virtualtoxlab – *in silico* prediction of the toxic potential of drugs and environmental chemicals: evaluation status and internet access protocol. *ALTEX 24(3)*, 153–61.

Veith GD (2006). Roles for QSAR in risk assessment. *ALTEX* 23(Sp. Issue): 369–72.

Velle S & Hal T (2004). Virtual frog dissection: reality check? [unpublished]. Cited in Cross TR and Cross VE (2004). Scalpel or mouse: a statistical comparison of real and virtual frog dissections. *Am Biol Teach* 66(6): 408–11.

Venning GR (1983). Identification of adverse reactions to new drugs. I: What have been the important adverse reactions since thalidomide? *Br Med J* 286: 199–202.

Villar D, Buck WB & Gonzalez JM (1998). Ibuprofen, aspirin and acetaminophen toxicosis and treatment in dogs and cats. *Vet Hum Toxicol* 40: 156–62.

Viscusi WK & Hakes JK (1998). Synthetic risks, risk potency, and carcinogen regulation. *J Policy Anal Manage* 17: 52–73.

Waldhalm SJ & Bushby PA (1996). Bringing information technology into the veterinary curriculum. *Semin Vet Med Surg (Small Anim)* 11(2): 96–9.

Wallenstein L & Snyder J (1952). Neurotoxic reaction to chloromycetin. *Ann Intern Med* 36: 1526–8.

Walsh PD *et al.* (2003). Catastrophic ape decline in western equatorial Africa. *Nature* 422(6932): 611–14.

Watters MPR & Goodman NW (1999). Comparison of basic methods in clinical studies and *in vitro* tissue and cell culture studies in three anaesthesia journals. *Br J Anaesth* 82: 295–8.

Weibel F *et al.* (1997). Genetically engineered cell lines: characterisation and applications in toxicity testing. The Report and Recommendations of ECVAM Workshop 26. *Altern Lab Anim* 25: 625–39.

White KK, Wheaton LG & Greene SA (1992). Curriculum change related to live animal use: a four-year surgical curriculum. *J Vet Med Educ* 19: 6–10.

White T (2007). *In Defense of Dolphins: The New Moral Frontier*. Malden, MA: Blackwell Publishing.

Whithear KG *et al.* (1994). Veterinary education in the era of information technology. *Aust Vet J* 71(4): 106–8.

Whitlow S, Bürgin H & Clemann N (2007). The embryonic Stem Cell test for the early selection of pharmaceutical compounds. *ALTEX* 24(1): 3–7.

Wilbourn J *et al.* (1986). Response of experimental animals to human carcinogens: an analysis based upon the IARC Monographs Programme. *Carcinogenesis* 7, 1853–63.

Wildman DE *et al.* (2003). Implications of natural selection in shaping 99.4% nonsynonymous DNA identity between humans and chimpanzees: enlarging genus *Homo*. *Proc Natl Acad Sci USA* 100(12): 7181–8.

Williams S, Butler C & Sontag MA (1999). Perceptions of fourth-year veterinary students about the human-animal bond in veterinary practice and in veterinary college curricula. *J Am Vet Med Assoc* 215(10): 1428–32.

Willmot M *et al.* (2005a). A systematic review of nitric oxide donors and L-arginine in experimental stroke; effects on infarct size and cerebral blood flow. *Nitric Oxide* 12: 141–9.

Willmot M *et al.* (2005b). Nitric oxide synthase inhibitors in experimental ischemic stroke and their effects on infarct size and cerebral blood flow: a systematic review. *Free Radic Biol Med* 39: 412–25.

Wilson JG *et al.* (1977). Comparative distribution and embryotoxicity of acetylsalicylic acid in pregnant rats and rhesus monkeys. *Toxicol Appl Pharmacol* 41: 67–78.

Wise S (2000). *Rattling the Cage: Toward Legal Rights for Animals.* Cambridge, MA: Perseus Publishing.

Wishart DS (2007). Improving early drug discovery through ADME modelling: an overview. *Drugs R D* 8(6): 349–62.

Wood JN *et al.* (2007). The perception of rational, goal-directed action in non-human primates. *Science* 317(5843): 1402–5.

Woodcock EA & Richardson R (2000). Effects of environmental enrichment on rate of contextual processing and discriminative ability in adult rats. *Neurobiol Learn Mem* 73: 1–10.

Worth AP & Balls M (eds) (2002). *Alternative (Non-Animal) Methods for Chemicals Testing: Current Status and Future Prospects. A Report Prepared by ECVAM and the ECVAM Working Group on Chemicals. Altern Lab Anim* 30(Suppl. 1): 1–125.

Young SS (1989). What is the proper experimental unit for long-term rodent studies? An examination of the NTP benzyl acetate study. *Toxicology* 54: 233–9.

Zhang H, Borman HD & Myhr BC (2004). Enhancement of the morphological transformation of Syrian hamster embryo (SHE) cells by reducing incubation time of the target cells. *Mutat Res* 548: 1–7.

Index